Tendu
(Diospyros Spps.)

Tendu
(Diospyros Spps.)

Vinod Kumar

2001

Daya Publishing House
Delhi - 110 035

ISBN 81-7035-256-8

Published by : **Daya Publishing House**
1123/74, Deva Ram Park
Tri Nagar, Delhi - 110 035
Phone : 7103999
Fax : (011) 7199029
e-mail : dayabooks@vsnl.com
website : www.dayabooks.com

Showroom : 4762-63/23, Ansari Road
New Delhi - 110 002

Laser Typesetting : Classic Computer Services
Delhi - 110 035

Printed at : Chawla Offset Printers
Delhi - 110 052

PRINTED IN INDIA

Dedicated to my mother

Preface

This book 'Tendu' is a compendium of valuable informations on various aspects of Tendụ including various studies made on 'Tendu'. The genus *Diospyros* 'Tendu' is the largest and most important of the Ebenacea and therefore the subject has been presented from every point of view. The Tendu bearing areas can be divided into three zones–dry, wet and intermediate. Seven species are found in the dry zone, thirteen in the wet zone and in the intermediate zone, where the rainfall varies from 175 cm to 200 cm, all the dry zone species except *D. melanoxylon* are found in addition to a couple of the 13 wet-zone species. It is a common belief that ebony or heartwood is dependent to a great extent on the nature of the soil. The proportion appears greatest in trees growing on rocky hillsides and decreasing as the quality of the soil improves, analogous in a way to the sandal, in which the proportion of scented wood also decreases with an improvement in the nature of the soil.

The most important part from the commercial point of view are the leaves of tendu (*Diospyros melanoxylon* Roxb.), used as wrappers of tobacco to produce bidi. It is a medium sized tree found throughout the dry forests of the plains, chiefly in Madhya Pradesh and Orissa, and to a lesser extent in Andhra Pradesh, Bihar, West Bengal, Maharashtra, Tamil Nadu, Gujarat, Rajasthan and Southern Uttar Pradesh. In certain parts of the country, leaves of 'apta' or 'ashitri' (*Bauhinia racemosa* Lam.) are also used as wrappers in the manufacture of bidi.

The annual collection of Tendu leaves in the country is estimated to exceed 300,000 tonnes, of which about 45 per cent comes from Madhya Pradesh alone. The annual collection in Orissa and Maharashtra has been estimated at 50,000 tonnes each and in

Andhra Pradesh at around 40,000 tonnes. The remaining production is shared by Bihar, Gujarat, Rajasthan, Uttar Pradesh and West Bengal; Bihar and Uttar Pradesh being relatively more important. Coppice shoots and root suckers of the tree produce good quality leaves and for this purpose, pollarding and pruning are practised in certain parts of the country. Pollarding / pruning is usually undertaken during the month of February / March and plucking usually starts about the middle of April and continues upto the end of May, extending some-times upto the first / second week of June. With the onset of rains, collection of leaves stops because of difficulty in drying. Utilisation of Tendu leaves is spread all over the country throughout the year. There are no assembling markets as such for Tendu leaves, and the important villages and towns in producing areas, where facilities for storage and transportation are available, usually serve as assembling centres. Madhya Pradesh, Maharashtra, Rajasthan, Orissa, Andhra Pradesh and Uttar Pradesh Governments have nationalised the trade in Tendu leaves. It is the experience of all the States that the nationalisation not only brought extra revenue to the States, but also ensured payment of reasonable minimum wages to the collectors.

As is the case with the production of Tendu leaves, so also utilisation is maximum in Madhya Pradesh and, therefore, this State is the most important bidi producing State in India. It is estimated that nearly 70 per cent of the Tendu leaves collected are used within the State itself. The remaining quantity is despatched mainly to West Bengal, Maharashtra, Karnataka and Uttar Pradesh. In Maharashtra, 65 to 70 per cent of the Tendu leaves produced is used in Andhra Pradesh and Tamil Nadu. The substantial quantity of best quality leaves obtained in Orissa are exported to other countries. The remaining quantity is utilised within the country, mainly in West Bengal and Maharashtra. Andhra Pradesh caters to about 60 to 70 per cent of the requirement of Kerala.

The Government of India had prescribed grade/standards for Tendu leaves and notified the Tendu (Bidi Wrapper) Leaf Grading and Marking Rules, 1963, under the provisions of the Agricultural Produce (Grading and Marking) Act. 1937. The Government of India also took recourse to the Indian Customs Act.1962 and prohibited with effect from August 15,1965 export of tendu leaves unless the same were graded and marked in accordance with the prescribed standards. The grade specifications are based on colour, texture, size, shape, body and conditions of the leaves and seven grades have been prescribed for the purpose, viz. Fancy, Extra special, Special, Good, Standard, Fair and General. Grading of Tendu leaves is beneficial to both the sellers and the manufacturers of bidis, but no proper attention is paid towards sorting and grading of Tendu leaves in general. Therefore, grading of tendu leaves as per 'Agmark' standards should be gradually extended to leaves which are consumed by the domestic industry. The Government of the States, where there is already state control over collection and disposal of Tendu leaves, should enforce grading of Tendu leaves prior to sale.

There is an urgent need to initiate research to correlate yield and quality of Tendu leaves in relation to pruning, pollarding of trees and other related operations. Control of pests and diseases, rotational working, regeneration, and a host of such problems remain to be tackled.

Tendu is an important name in the trade. Tendu leaves are exported out of India on a massive scale and fetch a handsome revenue. The various states such as Madhya Pradesh, Maharashtra, Uttar Pradesh, Andhra Pradesh, Orissa, Karnataka, etc. give employment to the large number of labourers for collection of Tendu leaves and wrapping of Bidi leaves. But the production of Tendu leaves is diminishing year by year. The reason being excessive harvesting, frequent fire and biotic interference by local villagers.

Tendu can not be propagated by vegetative means, therefore a comprehensive research is required to propagate it by vegetative means. The planting of Tendu by Nursery grown seedlings is also a challenging Job before the foresters. Large scale tendu plantation is necessary to give benefit to the poor villagers living in and around the forests.

Smoking is poisonous and injurious to health. It causes nervous dyspepsia, sleeplessness, gnawing of the stomach, Headache, Week eyes, red spots on the skins, throat irritation, asthma, bronchitis, heart failure, lung trouble, catarrh, melancholy, neurasthenia, loss of memory, impaired will power, rheumatism, lumbago, sciatica, neuritis, heartburn, torpid liver, loss of appetite, enervation, lassitude, lack of ambition and falling out of hair. Therefore, the smokers also have a right to know more and more about the tendu.

Tendu is repeatedly attacked by forest fire and hardly has a chance to grow without fire, therefore a Tendu tree takes a ugly shape. If not attacked by the fire, tendu grows to a beautiful and attractive thick crowned tree-but such type of trees are rarely seen.

I sincerely hope that this small book on tendu will give encouragement and interest to the foresters and trees planters in covering the suitable land with Tendu.

Amravati
2001

Vinod Kumar, IFS

Contents

Chapter 1
Introduction

Family Ebenaceae

The 'Tendu' is the largest and most important of the family Ebenaceae. The tendu bearing areas can be divided into three zones dry, wet and intermediate. Seven species of Tendu are found in the dry zone, thirteen in the wet zone and in the intermediate zone where the rainfall varies from 175 cm to 200 cm, all the dry zone species except *D. melanoxylon* are found in addition to a couple of the 13 wet-zone species. It is a common belief that ebony or heartwood is dependent to a great extent on the nature of the soil. The proportion appears greatest in trees growing in rocky hill sides and decreasing as the quality of the soil improves, analogous in a way to the sandal trees in which the proportion of scented wood also decreases with an improvement in the nature of the soil. The ebony of export is derived entirely from the genus *Diospyros*, although the *Dalbergia spp.* also provide marketable ebony.

Brief Description of Tree and Wood

'Tendu is a large tree attaining a height of 60 to 80 feet; sometimes seen as a shrub. It is native of "most woody mountainous countries viz India, Ceylon, Malabar, Coromandelet"- (Roxburgh). It is not present in Burma. *D. melanoxylon* mostly occurs in Western and Southern India whereas *D. tomentosa* might be said to be the Eastern and Northern Indian form. The "Flora of British India' retains these Roxburghian species as distinct but many Indian authors view them only as forms of one and the same species. The plant met with in the Ravi basin is *D. tomentosa* and not *D.*

melanoxylon. The *D. melanoxylon* is met with from Nilgiris, hills of Dindigul and Kurnool to Thane in Mumbai. It also occurs in Singhbhoom (W.B.). It is highly doubtful if any *Diospyros* yields a gum or resin, though the fruits of all would most probably afford a resinous juice.

The bark possesses astringent properties, and is used in decoction in Diarrhoea and Dyspepsia as a tonic. In a dilute form it is used as an astringent lotion for the eyes. Roxburgh says that the bark, powdered and mixed with pepper is given for dysentry.

The tree flowers in April and May and produces a fruit which, when ripe, is eaten by the natives. The fruits of *D. melanoxylon* is often confused with the fruits of *D. tomentosa*. Roxburgh's description of these two fruits may be given as under:

(a) *D. melanoxylon*–Fruit

Berry round, of the size of a small apple. Yellow, pulpy, seeds as many as 8, immersed in pulp. Kidney shaped. Sharp on the inner edge.

(b) *D. tomentosa*–Fruit

Berry ovate, as large as pigeon's egg, covered with a smooth hard bark, which becomes yellow when ripe and is filled with a soft, yellow edible fruit. Dauglass mentions a variety without a stone which is cultivated in central Asia highlands. The seedless tendu tree was found in the S. Etapalli range of Bhamragarh division of Maharashtra. In that tree the fruits growing in the middle and top of the crown are seedless but on the same tree the fruits growing at the lower of the crown are having seeds of smaller size. The whole tree is not seedless. It is a common expression among the tribals that they live on Mohwa and Tendu.

The wood is hard of a light pink colour with irregular shaped masses of black ebony in the center. The ebony is jet black with purple streaks, extremely hard, pores and medullary rays difficult to distinguish. It is used for buildings, shoulder, poles and carriage shafts and the ebony for all purposes of furnishing work and carving.

According to Talbot — "The wood is used for building and is fairly durable. Blocks of ebony are found in the center of old trees from 12 to 18 inches in diameter, and on an average weighing from

75 to 80 pounds per cubic feet. Only the heartwood of old trees contain ebony, and even that is streaked with dull yellow lines. The wood, though strong, tough, and fairly durable is not held in much esteem. A seasoned cubic foot weighs 50 to 70 pounds.

(c) Botanical Description

The exact botanical description of *Diospyros melanoxylon*: Roxb plant has been a matter of controversy on account of the great variability exhibited by the leaf. A modified botanical description of the plant was given by Rathore (1972) recognizing four basic leaf forms on the basis of shapes namely: Table 1.1

Ellipsoidally lanceolate	-	Form A
Ovate	-	Form B
Elliptic	-	Form C
Orbicular	-	Form D

Several writers of Indian flora found it difficult to define the precise limitations of the species *D. melanoxylon* on account of leaf variability. Cooke (1908) in his work "The Flora of the Presidency of Bombay refers to the separate identification by Clarcke of two leaf specimens collected by Dr. Ritchie from the hills north of Belgaum, at two different seasons of the year but possibly from the same tree as both specimens bear the same herbarium number (kew 1108 and 1108/2). Of these one has been named by Clarcke as *D. tupru* and other *D. melanoxylon*. Troup (1921) has differentiated *D. melanoxylon* from *D. tomentosa* on the basis of narrower leaves, acute apex and base and with secondary nerves raised in comparison to rounded or obtuse leaves of *D. tomentosa*. However silviculturally both the species were regarded as one *D. melanoxylon*, by Troup (1921). Hooker (1882) described the leaf of *D. melanoxylon* as elliptic but narrower at both ends, of *D. tomentosa* as large and ovate and those of *D. tupru* as smaller, elliptic and which are obtuse at both ends. All these three species were merged in. *D. tomentosa* by Brandis (1874) and in *D. melanoxylon* by Gamble (1922). The situation is really complex as all the four leaf forms together with their inter grades are found in the same plant. However, individual plants show a predominance of any one type (Rathore, 1972).

Hooker (1882), had differentiated *D. melanoxylon D. tomentosa* and *D. tupru* on the basis of leaf shape which is elliptic with acute apex and base angles in the case of *D. melanoxylon*, ovate in *D.*

tomentosa and elliptic but with obtuse apex and base angles in the case of *D. tupru*. According to Hooker (1882), the leaf of *D.tupru* rarely exceeds the length of 8.8 cm and the leaves of *D. tomentosa* and *D. melanoxylon* show lengths of about 20 cm and 10 cm respectively. The leaf form B is the only ovate form and also shows the highest value for length (15.92 ± 1.22). This should be the *D. tomentosa* of Hooker (1882). However the name tomentosa may not appear appropriate as the species shows the lowest values of epidermal hairs. Between the two elliptic forms i.e. A and C, segregation of *D. tupru* and *D. melanoxylon* of Hooker (1882) is not difficult. Leaf length is one important criteria which is 10.74 ± 1.58 in the form A and 7.75 ± 2.64 in the form C. The lamina apex and base angles are also acute in the form A and obtuse in C. Thus leaf form A is *D. melanoxylon* and form C is *D. tupru* of Hooker (1882). There is no reference to orbicular leaves in either of these three 'species' in flora records.

Since all these leaf forms are found in individual tree (Rothore-1968), they represent intravariations of a single species rather than leaves of three distinct species. This also finds support in the stomatal index which shows a mean value of 16.37. According to Wallis (1957) stomatal index remains more or less constant for a given species.

Leaf Area of Tendu Leaf

According to Shri S. K. Puri (Indian Forester May-1964) the leaf area of a tendu leaf can be determined by a simple formula. If a = length of tendu leaf in cm; b = width of tendu leaf in cm; then y = area of the leaf in square cm will be

Y = 0.7590 a. b. (for Tikamgarh division)

But in my opinon this equation will not fit exactly for all the places. The leaves of tendu resemble very much to an ellipse and the equation for the area of tendu leaf needs to be formulated for each place separately.

Tendu or kendu is very common in the deciduous forests of Madhya Pradesh, Orissa, Bihar, Uttar Pradesh, West Bengal, Andhra Pradesh, Maharashtra and Gujarat. The other names of tendu leaves are kendu, timru, temburni, balai or tumki in different states. Generally two types of tendu leaves commonly occurring in Indian forests are *Diospyros melanoxylon* and *Diospyros tomentosa*. The Tendu leaves are used as wrappers for the manufacture of 'Bidis' (substitute for Cigarettes) which are mostly smoked by the lower middle class and poor people.

A 'Bidi' is prepared by putting tobacco or 'Jarda' in a piece of tendu leaf, then rolling it, giving a conical shape, completely folding in and closing at broad end; partly closed at the other narrow flat end. The rolled tendu leaf is tied with fine cotton thread and then baked to the desired extent.

The semi – matured juvenile tendu leaves are collected from pruned or cut back tendu plants or tendu bushes. The leaves are dried and processed before using as bidi wrappers.

The leaves of other species which are also used as bidi wrappers are:

Kanchan or Apta	-	*Bauhinia recemosa, Bauhinia Vahlii*
Kurai	-	*Hollarhena antidysentrica*
Sal	-	*Shorea robusta*
Banana	-	*Musa paradiscioca*

Out of these Kanchan (Apta or Ashiri) are collected commercially to a small extent in a few localized pockets in Dangs, Surendranagar and Junagarh districts of Gujarat, Ratnagiri, Nasik, Pune and Thane in Maharashtra as well as Moradabad in Uttar Pradesh.

The use of Kanchan leaves as bidi wrappers is very limited due to the reasons that:

(a) Kanchan leaves can not stand baking or hot air treatment.
(b) Cost of rolling of Kanchan leaves are higher as compared to tendu leaves.

Diospyros melanoxylon (Roxb) 'Tendu'– Why Used for Bidi

'Tendu' is extensively used in manufacture of 'Bidi' and its leaves fetch substantial amounts of revenue to the Forest Department of States like Madhya Pradesh, Maharashtra, Andhra Pradesh and Orissa. At present India is the biggest 'Bidi' producer in the world and also the biggest consumer of this commodity; about 88,000 tonnes of Bidis are being consumed in India annually.

The flavor, texture and workability of tendu leaves are unmatchable and preferred for the following reasons:

(a) Thin and fine texture of leaves make it pliable when moistened for easy rolling of 'Bidis'.

(b) Tendu leaves possess hygroscopic nature i.e. they become extraordinarily soft after moistening which renders 'Bidi' rolling easier without breaking.

(c) Tendu leaves have characteristic aroma which go with the Bidi–tobacco producing of desired flavour while smoking.

(d) Slow and constant burning of 'Bidis' without getting extinguished once lighted. But the practical experiences show that this is not true for very old stored 'Bidis'.

(e) Tendu leaves dry up quickly without developing wrinkles, cracks, and fractures. The bidis so made can with-sand baking and hot air treatment.

(f) Capacity to with sand storage without deterioration of quality for nearly five years.

(g) Large scale abundant availability for commercial 'Bidi' productions.

For the above reasons the leaves of *Diospyros melanoxylon* are extensively used for the manufacture of 'Bidis'.

Chapter 2
Description of Tendu

The commonly occurring tendu leaves in the Indian forests are *Diospyros melanoxylon* and *Diospyros tomentosa* belonging to the family Ebenaceae. Tendu is called by different names.

Vernacular	-	Tendu, Kendu, Temru
Hindi	-	Tendu
M.P.	-	Tendu,Timburni, Temru,Tumri, Tumburni
Marathi	-	Temburni Timru,
Gujarati	-	Timru, Balai
Bengali	-	Terel, Kiril, Kendhu
Oriya	-	Kendu, Kend
Telgu	-	Tumki, Tummuk
Karnataka	-	Tembhurni.

Diospyros melanoxylon and *Diospyros tomentosa* are closely allied but are considered to be two important separate species for bidi manufacturing in India.

(A) Diospyros melanoxylon (Fig.1)

This is a small and bushy or a large straight tree with grey or rusty *tomentosa* shoots. Bark is grey and black mixed exfoliating in irregular straites. Blaze is chrome yellowish white, leaves nearly opposite, short and petioled, small oblong to elliptic, narrowed at both ends. The old leaves are thicker, rather brittle, coriacious rugoes appearance above and *tomentosa* to glabarous beneath. Nerves conspicuous faint to strong, reticulations raised above young leaves are villous. Size of leaves 10 to 20 cm. long, 2.5 to 7.5 cm broad. The leaves are mostly parabolic in shape, rather rough shining above usually and greenish below. Coppiced leaves are larger and thinner therefore best suited for 'bidi' manufacture. Deciduous towards late February, occasionally evergreen in damp localities.

In its natural habitat the maximum shade temperature varies from - 1°C to 13°C, the normal annual rainfall varies from 500 mm to 1500 mm. Seedlings and young poles stand moderate shade but for later development, more light is required. The seedlings are resistant to frost and drought but susceptible to excessively moist conditions. The tree coppices well, but the coppice shoots grow slowly, it pollards better, though the growth of pollard-shoots is also slow. It reproduces easily through root suckers.

Male Flowers: Nearly 4 to 6 (or even more) in pendunceled 3-12 flowered, villous, simple or branched, often dripping comes, bands narrowly ellipsoid, 6 -7 mm long with tubular calyx (half as long as the silky corolla or more), corolla after opening 5 to 6 mm long. Tubular funnel shaped with suberect small obtuse lobes.

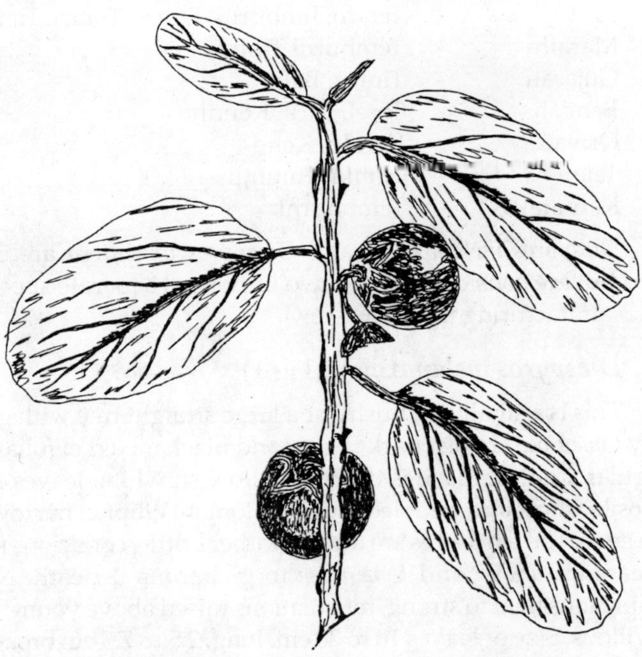

Fig. 1: Diospyros melanoxylon

Female Flowers: Solitary axillary or from leaf scars, sub sessile or on short thick peduncle, calyx broad with edges curved 4 to 5 angled, corolla 7.5 to 10 mm similar in shape to male flower, staminodes 8 to 10 ovary densely hairy 4 to 8 celled, styles 3 stigma bifid.

Fruit: Subsessile small, globose to ovoid, dull yellow when ripe, 2.5 to 4 cm diameters.

Seeds: Upto 8 kidney shaped and compressed. Testa rugose shining.

Flowering: May - on new shoots.

Fruiting: Fruit ripens in May, edible.

(B) Diospyros tomentosa

The descriptions of *Diospyros tomentosa* is similar to that of *Diospyros melanoxylon*. But the following characteristics are noteworthy to distinguish it:

(i) **Bark:** Very thick, black and exfoliating in rectangular pieces, blaze is yellowish white, darkens after exposure to air.

(ii) **Leaves:** Sub-opposite to alternate, long petioled, wide, oviate with round base, more or less hair or tomatoes beneath. Nerves strong and reticulations impressed. Leaves on mature trees 10 to 20 cm x 5 to 7.5 cm in size. Leaves deciduous by end of December or early January.

(iii) **Female flowers:** Styles two - fruits - globose, large, shining yellow fruit, seeds up to 5 or less , not much compressed.

(iv) **Identification of leaves:** The leaves of two species can be distinguished on the basis of their leaf structure as follows:

D. melanoxylon	*D. tomentosa*
Leaves elliptic, narrowed at both ends, tip obtuse or rounded or sub acute, base usually acute, in form, base and apex often acute secondary nerves raised on upper surface.	Leaves ovate, base rounded, obtuse base, nerves impressed on upper surface.

(C) ***D. paniculata*** (Fig.2)

Sanskrit - Thinduka
Tamil - Kurunduvati
Malyalam - Kari, Karivella, Illakatta

A moderate sized handsome tree attaining height of 15 meter and a diameter of 38 cm occurring in the evergreen rain forests of western ghats, ascending up to 1000 meter. The fruits are green and avoid, about 2.5 cm long.

The wood is whitish grey, occasionally with narrow stripes of black. It does not yield a black heart wood. It is soft and moderately heavy (730.3 kg/cu. m) and is said to be suitable for match boxes.

The leaves of the tree are used as fish poison. Dried and powdered fruits are applied to heal burns. A decoction of the fruit is used in gonorrhoea, biliousness and blood poisoning, and powdered bark is used for rheumatism and ulcers. It is found in the forests near the Nilkund and Gairsoppa ghats.

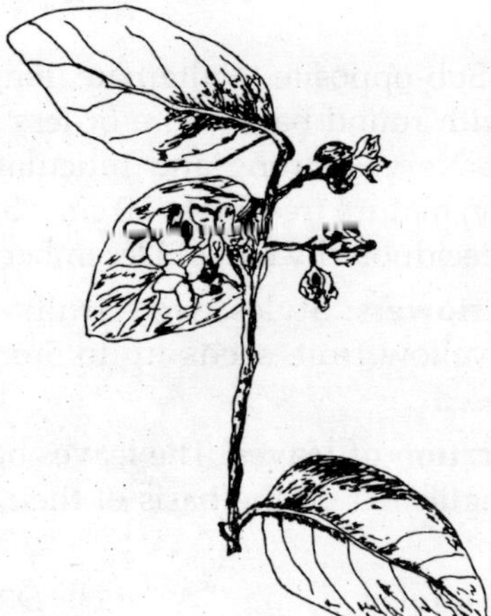

Fig. 2: Diospyros paniculata

(D) ***D. peregria*** *(syn. D. embryopteris ; D. malabarica)* (Fig.3)

Desr. GAUB PERSIMMON

Sanskrit - Tinduka, Krishnasara, virupaka
Hindi - Gab,Kalatendu, makurkendi
Bengali - Gab, makurkendi, tendu

Marathi	-	Timburni, temburni
Telegu	-	Tinduki, gabu
Tamil	-	Kattatti, Kavikattai, tumbi
Kannada	-	Holetupare, kusharta,
Malayalam	-	Panchi vananji
Uriya	-	Dhusarokendu, kendu

A moderate sized evergreen tree with short straight, often fluted trunk and spreading branches found practically throughout India in shady wet places and near streams. It bears dark green foliage and large velvety fruits, and is cultivated for ornament.

The tree is shade bearer and in the natural state, young plants develop freely under fairly heavy canopy. It thrives on moist soils and on ordinary loam provided it is not too dry.

The fruits fall to the ground from June or July onwards and under favourable conditions, the seeds germinate during the rainy season.

Gaub parsimmon fruit is a special berry, as big as a middle-sized apple, with a leathery rind, containing 4-8 seeds embedded in a viscid glutinous pulp. It is yellow when ripe and is covered with a rusty, easily detachable scurfiness. A good tree produces about 4,000 fruits in a season. Fully ripe fruits have a mawkish sweet taste and are edible, ripe fruits are resistant to insect attack.

Analysis of dried and powdered fruits gave the following values;–ether extract 1.2; alcohol extract 12.4, water extract 7.5, albuininous matter 12.1, organic residues 61.9; and ash 4.9%. The fruits contain a high percentage of pectin (50%) and being cheap and abundant, constitute a promising source of sizing material. Unripe fruits are rich in tannin and are employed for tanning hides and dyeing cloth. Pulped fruit is used as a preservative for fishing nets and as glue for book binding; boiled with or without powdered charcoal, it is used for paving bottoms of boots.

The fruit and the stem bark (tannin content; fruits, 15% and bark, 12 %), possess astringent properties. The unripe fruit is acrid, bitter and oleaginous. An infusion of the fruit is used as gargle in alpathae and sore throat. The juice forms a useful application for wounds and ulcers. The bark is used in dysentery and intermittent fevers. The oil extracted from the seeds is used for dysentery and diarrhoea.

The wood of gaub persimmon is grayish in colour, close grained, moderately hard and heavy (wt.76.2-77.8 kg) per (cu. m). It is some times employed for building and constructions and in boat making. The wood yields charcoal (ash, 4.9%) suitable for producer gas plant. The leaf and fruit has been shown in (fig.3).

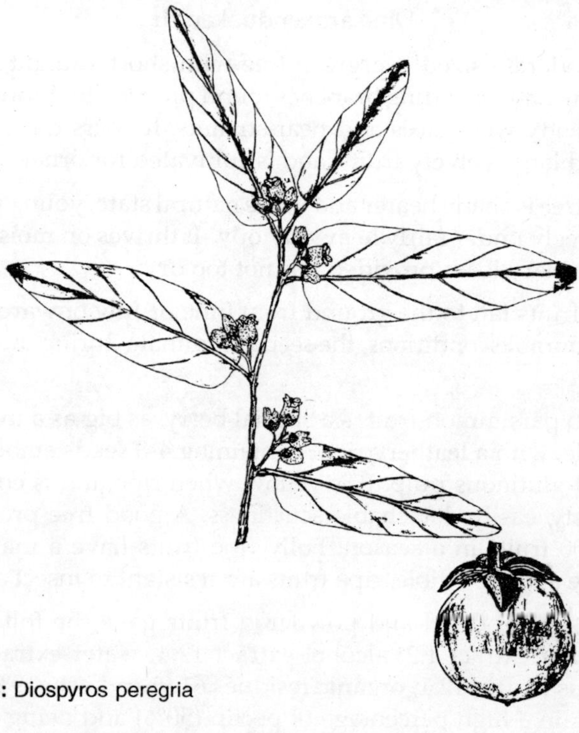

Fig. 3: Diospyros peregria

(E) *D. sylvatica*

Telegu - Gadaluti, gada
Kannada - Akkasarali, bilisarali
Oriya - Khalijya, Modhurokhalijya, Kalicha, Kanchia

A medium sized, sometimes buttressed tree attaining a height of 18 meter and a girth of 105 meter distributed in Orissa, South India and Ceylon. Bark coloured, thin, smooth, thinner bark yellow. Leaves 7.5 to 12.5 by 2.5 to 6 cm elliptic, acute or acuminate, membranous, glabrous and shining above, paler beneath, base usually acute. Flowers white, fragrant fruit globular, 2-4 cm diameter, edible, dark green seeds 2-8 oblong. 1.25 cm long, shining, brown albumen ruminate.

The wood is grey with black streaks and irregular black patches at the center. It is moderately hard and heavy 793.7 kg/cum. It is used for fancywork.

(F) *D. condolleana*

A small tree 12 meter high, young shoots pubescent.

Leaves 10-15 by 4-6 cm elliptic-oblong, globarous on both sides when mature, base narrowed or rounded, midrib strong beneath, lateral veins nerves and veins inconspicuous. Flowers white in short axillary 2-10 flowered cymes, seeds 3-4 usually compressed; tests thin, brown; albumen ruminate, occurs throughout the Konkan and Kanara in evergreen forests.

(G) *D. crumenata* (Fig.4)

A very large tree with cylindrical trunk reaching 33-50 metre high (Talbot); bark thin, scaly. Leaves 6-13.5 cm by 2.5-5 cm, elliptic oblong, base rounded or slightly narrowed. Flowers white, tubular, about 1.25 cm long, lobes 4 seeds, black shining 2.5 cm. It is confined to Kanara where it is common in the evergreen forests.

(H) *D. oocarpa*

A small or moderate sized tree with thin smooth bark; young shoots pubescent. Leaves alternate 8.5-11 by 3.25 -6 cm ovate-elliptic, glabarous and shining, base usually rounded, nerves and veins inconspicuous. Flowers white, fruit 2-3 cm long, cylindrical when young, ellipsoid when ripe, rounded at the top, 6-celled. Seed oblong, albumen not ruminate. Found in the evergreen forests of N. Kanara, common on the ghats from Karwar to Ainshi (Talbot).

(I) *D. pruriens*

A small tree, bark dark coloured scaly. Leaves 5-10 by 2-4 cm, elliptic oblong, base usually rounded, sometimes subcordate. Flowers white, ovary 4-celled, fruit ovoid- conical 2-3 cm long 4-celled, 4 seeded. Found in the evergreen forests of the southern ghats of N. Kanara (Talbot).

(J) *D. ebenum*

A large or moderate sized tree with sub erect trunk and leafy head; bark thin, grey; branchlets globarous. Leaves 5-15 by 2-5 cm glabrous, elliptic oblong, or ablong-lanceolate, bright green and

Fig. 4: Diospyros crumenata

shining above. Flowers pale greenish-yellow 1 cm long before expansion, lobes 4. Seeds 3-8, dark black. 1-cm long albumen not ruminate. Found in the evergreen forests of Kanara.

The wood furnishes the ebony used for wood carving in Kumpta and Honavar - N. Kanara. The quality is good but large pieces are rare.

(K) *D. microphylla* (Fig.5)

A large tree 25-35 meter high (Talbot), with a straight trunk, unbranched for a considerable height, buttressed at the base; bark dark, scaly, young branches verticillate, the older branches glabarous. Leaves 2-5 cm 1-2.25 cm, ovate elliptic, acute or obtusely acuminate, glabarous above, base tapering. Flowers white 0.25 to 0.35 cm long, 4 lobes rounded, seed 1 (rarely 2); albumen not ruminate. Found in the evergreen forests of N. Kanara.

(L) *D. Cordifolia*

A small tree with strong often branched thorns scattered over the trunk and larger branches. Leaves 3.75-7 cm by 0.8-1.8 cm,

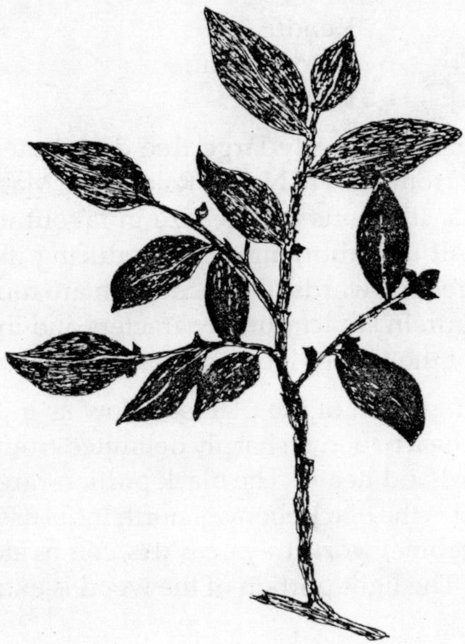

Fig. 5: Diospyros microphylla

oblong, obtuse, velvety on both sides, base rounded. Flowers white 1.25 cm long. Fruit 1.25-2.5 cm in diameter, globose, yellow when ripe. It is tolerably common in N. Kanara.

(M) *D. chloroxylon*

A small tree bark deeply cracked; branch lets tomentose. Leaves alternate 2.5-6 by 1.25-4 cm long, elliptic, obovate or elliptic-oblong, nerves inconspicuous. Flowers white 4-10 together 0.1-0.2 cm long. Fruits globose 1.25 cm in diameter, seeds 2-3, albumen not ruminate. Found in Nasik–Peint road about 10 km. Common about Surat on the north side of the river. Also found in Central and Southern India , Orissa and Karnataka.

(N) *D. tomentosa*

(Nepal–Ebony persimmon)

Hindi	-	Tendu, kendu, temru, tumri, tumki
Bengali	-	Kend, kyon
Telegu	-	Citta tumiki, man citumiki, tumiki
Tamil	-	Tumbi
Kannada	-	Timburani, tumari, tindura

Uriya - Kendu
Punjabi - Tendu, kinnu
Trade - Ebony

A small or occasionally large tree distributed in the sub-Himalayan tract from Ravi to Nepal, Rajputana, Madhya Pradesh, Bihar and Orissa. It attains a large size in favourable situations, but remains small in scrub jungles, reproducing itself from root suckers. Large trees upwards of 180 cm girth are found in Kangra and Chota Nagpur. In silvicultural characters and in the structure and properties of the timber it closely resembles *D. melanoxylon*.

The rate of growth of the plants is slow as in the case of *D. melanoxylon*, the heartwood is sharply delimited from the sapwood and is black, hard and heavy. The black portion rarely exceeds 15 cm in diameter. It is the black ebony of north India used for carving, picture frames, cabinet work, trays, caskets, combs etc. It is suitable for brush backs. The light portion of the wood is extremely strong and elastic.

(O) *D. embryospteris* (Fig.6)

A small or middle sized tree glabarous except the very young parts and the inflorescence; branches numerous, spreading, forming a dense shady head; bark smooth, thick, blackish, flaking off in large pieces. Leaves variable in size numerous, 10-18 by

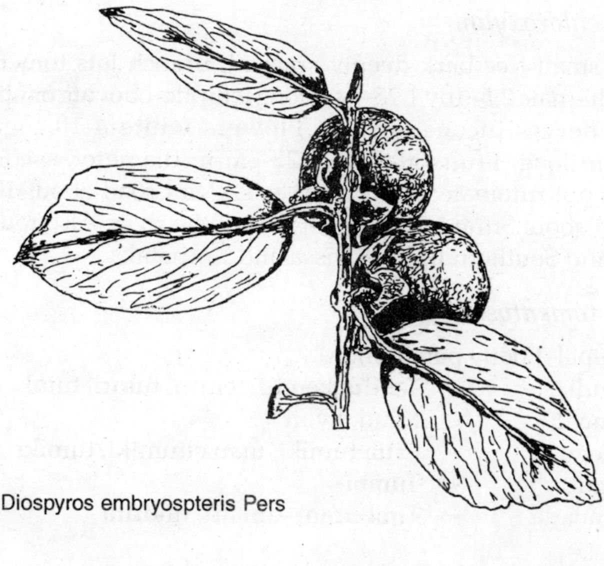

Fig. 6: Diospyros embryospteris Pers

2.5-8 cm oblong or oblong lanceolate, glabrous and shining, the veins prominent in dried specimens, obscure in fresh ones, base usually tapering. Flowers white, 4 lobed. Fruit large 3.5-7.5 cm in diameter, usually solitary, 8-celled pericarp thin, pulp viscid and glutinous, seeds 4-8 flattened, smooth, reddish brown. It is found near creeks and back waters along the coasts of the Konkans and N. Kanara, also in evergreen forests of the ghats and in the beds of rivers and streams (Talbot). It occurs in major parts of India; also in Ceylon and Siam.

The unripe fruit is very astringent and an extract is employed in medicine. It is also used as preservative for fishing nets. The viscid pulp is used as glue for bookbinding.

Silvicultural Characteristics of Tendu

Tendu is regarded as one of the hardiest trees, tendu tolerates shade well in its early stages but requires more light for its full development. It resists drought and frost extremely well. It is immune to grazing and browsing but tender buds are browsed occasionally. It grows on poorest soils and even on fairly heavy clays although it does not attain large size. It coppices and produces root suckers freely. Its rate of growth is very slow. Tendu develops a long tap root at an early stage.

Tendu leaves shed leaves or defoliate from December to February normally and new flush of leaves comes in March - April depending upon species, weather conditions and water table of the site. Tendu also sends forth second flush of new leaves (Kartika) again between October and November. Normally Kartika leaves are not used for bidi manufacturing.

The new leaves that are produced from March to May (and even in June-July) by the method of pruning or coppicing of small plants and bushes are usually collected for 'Bidi' wrapping. After second week of June the environmental conditions are not favourable for Tendu leaves processing. Further the 'Kartika' leaves are comparatively thicker and impart more acrid taste in smoking. Kartika leaves are difficult to dry during winter due to dew, occasional showers and lack of strong sun. There is no demand for Kartika leaves.

The phenomenon of leaf shedding (deciduation) and leafing enables the plant to grow continuously till such period (senescence) when the total quantity of new leaves on the tree gets reduced year after year. Schimper (1903) described an extreme in spps. *Urostigma glabellum* where leaves were shed within a day thereby a tree that appeared quite verdant in the morning became bare of leaves in the evening. After ten days the tree again became green. Such a rapid development of leaves is called by early botanists as 'out pouring' of leaves (Taylor 1970). The age of individual leaves of tendu is about ten months. The rate of growth of tendu leaf is maximum when leaf size is minimum and decreases slowly as the leaf attains normal size. The growth pattern follows a sigmoid curve. During leaf development the length to breadth ratio of the leaf almost remains constant i.e. the leaf develops without change in shape (homoblastic development). The red colour in the leaves is due to anthocynin pigment.

(a) Rainfall and Temperature

The normal annual rain fall in tendu growing areas varies from 500 mm to 1500 mm, maximum shade temperature varies from 40°C to 48°C whereas minimum temperature ranges from 1°C to 13°C (Dharet all 1989 and Troup 1921, 1986).

(b) Reproduction

Tendu seeds geminate well. It requires hot water treatment. Boiling water treatment is injurious, even cold water treatment for 48 hours has proved good. The rate of growth of Tendu plants in the nursery or in the field is very slow. In the field it requires manures and phosphatic fertilizers.

Natural reproduction of tendu is not common. The seeds which passes through the alimentary canal of goats, monkeys, deers and other animals germinate very well and therefore the natural regeneration of Tendu is noticed in clusters in the forests.

Natural regeneration by root suckers is very prominent and is considerably enhanced by: (a) Fire and (b) Ploughing.

The fire is of common occurrence in tendu bearing forests. Fire although enhances natural regeneration of Tendu by root suckers considerably but it damages other regeneration. In teak bearing areas the teak seeds fallen on the ground are treated by the fire and profuse natural regeneration of teak is observed which get established provided there is no fire in the next two years.

(c) Distribution

Tendu occurs as a middle story associate in tropical dry and moist deciduous forests. It is also found in the tropical semi-evergreen forests and in sub tropical sal forests. It has a wide range of distribution, extending from central pennisular India, throughout Malwa in the West to Orissa and Bihar in the East and gangetic plains in the north or Andhra Pradesh in the South. Tendu is commonly found in the forests of Madhya Pradesh, Orissa, Bihar, parts of Maharashtra, Uttar Pradesh and Andhra Pradesh; it is also found in certain tracts of Gujarat and West Bengal.

Tendu occurs over a wide range of variety of geological formations, soil types, topographical features and climatic regions. In forest areas big trees, often 2 m to 3 m girth are seen where conditions are favourable but in low quality forests the girth seldom exceeds one meter and height 15 meters. But in the forest areas near the villages the growth is arrested to bushy form due to repeated hacking by the villagers who use Tendu as poles, firewood; fencing material and for various other purposes.

Tendu not only grows in sheltered valleys with good loamy to clayey soils but also on denuded and poor soils which are of little value for raising tree crop.

(d) *Diospyros melanoxylon* Roxb

It occurs in dry mixed forests of peninsular India. It is a conspicuous constituent species in all the three major types of forests i.e. teak forest, sal forests and mixed forests. At places, it forms almost pure consociations of limited extent. This species is abundantly found in M.P, Maharashtra, Orissa and Andhra Pradesh; other important tendu producing states are Rajasthan, Uttar Pradesh and Gujarat. It sparsely occurs in West Bengal, Tamil Nadu, Karnataka and Kerala.

(e) Site Features

Tendu occurs over a wide variety of geological formations, soil types, topographical features and climatic regions. As per Troup (1921) tendu is important in clothing dry poor ground and has a wonderful hardiness in surviving maltreatment. It reaches its best development on metamorphic rocks but tolerates a large variety of soils. It also occurs on denuded and detrital soils often not of much value for raising tree crops of timber value (Ghosh et al 1976). However, it avoids clayey and water logged situations. It is commonly observed around forest margins and revenue wastelands where ravages of hacking, grazing and fire are common.

Tendu trees if felled, produces a number of coppice shoots from the stump. The superficial roots of tendu give out root suckers naturally and this process is augmented if these roots are injured or severed. After repeated hacking the coppice shoots as well as root suckers adopt a bushy form. Tendu reproduces itself copiously from root suckers on newly cultivated lands. Tendu is a very good coppicer and coppices freely. Due to repeated ploughing of land or hacking the tendu bushes are formed in open land around villages and as undergrowth inside the forests. Tendu in bushy form tends to dominate because it is fire hardy, is rarely browsed by animals and has got extraordinary powers of proliferation by sending out coppice shoots and root suckers. The tendu trees around the human habitation are subjected to ravages of hacking, grazing and fires, due to these reasons the tendu trees become ugly. Tendu tree is a very handsome tree full of thick foliage if not disturbed by hacking, browsing and fire but such trees are very rare. Such beautiful tendu trees can be found far away from human habitation from where no Tendu leaves were ever collected and there was never any fire in the area.

Regeneration of Tendu

There are two methods of regeneration.

(a) Natural Regeneration

The natural regeneration of Tendu is plentiful around trees due to germination of seed and also due to reproduction from the root suckers. The seeds passed through the animals (deers, monkeys, goats etc) germinate quickly.

Germination begins early in the rainy season and continues throughout leaving some seeds dormant until the second rainy season, germination is greatly favoured if the seeds get covered with soil as this protects the radicals from insects and other unfavourable conditions. However, the profusion and tenacity of sucker reproduction is in itself enormous to ensure propagation and survival of this species.

Ploughing of the Tendu area shows remarkable result of natural regeneration by root suckers. A characteristic feature of this species is the extensive production of root suckers and their hardiness and immunity from damage by animals. In a study to know further about the curious creeping habit of the young tap root, Rathore (1971) found that the main tap root which goes down into the soil to a depth of 1.5 m, produces root suckers upto a depth of about 1 m.

(b) Artificial Regeneration

The Tendu seeds do not germinate quickly. The Tendu seeds require hot water treatment of 80°C (allow it to cool naturally). The boiling water treatment is injurious to it. The soil mixture for raising Tendu in polybags should contain soil, sand and FYM in the ratio 1:1:1. The nursery does not require shade. The seeds must be sown during early May in polybags. The germination start after 20 days. The growth of plants is very slow. By the end of July the plants can be taken to the field for planting in pits at an spacing of 3 m x 3 m or 4 m x 4 m. The pits of size 30 cm x 30 cm x 30 cm is made for planting tendu. While half filling the pits sufficient FYM nearly half of the pit and also the potassium fertilizer 20 gm/pit must be added in the soil otherwise the Tendu plants will suffer from 'die - back' and may not grow at all. Therefore the planting Tendu becomes costlier. Tendu grows very well in soils having pH between 6.5 to 7.5. For soils having pH of 8 or more the addition of FYM in the pits is a must.

Two weedings in the first year and two weedings in the second year are sufficient for Tendu plants. Just after weeding Nitrogen fertilizer 10 gm/plant should also be added after each weeding. But addition of Potassium fertilizer will not do any good to the plants. Addition of fertilizer helps in the production of better sized tendu leaves.

Uses of Tendu Wood and Leaves

Tendu wood consists of two portions, outer part or sap wood is whitish brown or reddish brown whereas the central wood or heartwood or 'ghabha' is perfectly black. The heartwood portion is rarely of utilizable size. Only in big trees black heartwood of utilizable size is found. The black heartwood is heavy, hard and comparatively brittle. The heartwood is used for small furnitures, carving and ornamental purposes. It polishes well but generally it cracks if nail is inserted in the timber therefore its use as small furniture is also limited. The sap wood is suitable for tool handles, jute shutters and bobbins. Tendu poles are widely used for fencing and construction of huts. Tendu wood when burnt emits sparks known as Chirchiri. Tendu fruits are edible-they are very sweet just like 'Chiku.' The fresh and juvenile leaves are used for manufacturing bidis. Good quality and large sized 'bidis' are exported out side of India and give large amount of revenue. Tendu fetches large amount of revenue to the forest department.

Chapter 3

Artificial Regeneration of Tendu

(a) Seed Collection and Storage

The best time for seed collection is from April to June when ripe fruits are available. In tendu the phenomenon of alternate good and bad seed years exist. Seeds are easily extracted after removing the edible pulp from ripe fruits, and dried for about a week in the sun. For getting 1 kg of airdry seed 8 kg of freshly collected ripe fruits are required.

Seeds can be stored upto one year at the most. Fertility tests carried out at Dehradun on seeds of different origins and stores for one year showed loss of viability varying from 40 to 90% (Dent: 1948). One year old seeds from Markanda in Maharashtra gave only 25% germination. Experiments conducted in Pakistan show that seeds sown 7 months after collection gave 30% germination (Qadri : 1962). Fresh seed is therefore, always preferable to secure a good germination per cent.

(b) Seed Weight

Sen Gupta (1937) reports that a kilogram of seed obtained from Travancore contains 1060 seeds. According to Troup (1921) there are 880 to 1410 seeds in the same one kg weight. Reports from Maharashtra indicate that on air drying 1 kg contains about 1200 seeds.

In a study conducted in Maharashtra it was found that the number of well developed seeds per kg of weight varies according to the size of fruits as given in Table 3.1.

Table 3.1: Relation between no. of fruits / seeds and weight

Size of fruits cm	No. of fresh fruits per kg	No. of seeds	Weight of seeds gm
4.80 x 4.43	32	115	255
3.73 x 3.36	41	127	260
3.36 x 3.00	93	190	245

(c) Nursery Technique

Because of the large size of the tap root, transplanting from nursery is attended with much risk. Stump planting is also not successful. Best method is to sow seeds in long narrow bamboo baskets or other containers immediately after collection of seeds. Pure sand or light sandy loam has been found to be a better germination medium than ordinary soil or clayey loam. After sowing, the seeds should be covered with a thin layer of earth and regularly watered which should be done carefully, keeping in mind that seedlings are sensitive to excessive moisture the plants raised in bamboo baskets should be planted out intact at beginning of rains in 30 cm pits at an escapement of 3 m x 3 m.

(d) Formation of stands by direct sowing

The seeds are sown in April-May at suitable intervals. For the proper development of seedlings frequent soil working and weeding is essential. During the first year or two the seedling development is mainly underground and long tap root is formed rapidly attaining a length of 30 cm or more within a few weeks and a length of 60 cm by the end of the second season. Growth above ground is comparatively slow, a maximum height of 7.5 or 10 cm being attained by the end of the first season. Cereal crops can be grown in between the lines as they are shallow rooted and do not compete with the tendu plants. Leaves suitable for plucking can be obtained from such plantations after 5 years.

(e) Germination trial

Studies on germination of tendu have been carried out in Maharashtra and Pakistan. The observations are given in Table 3.2.

Table 3.2

Locality	Period of germination commencement	Germination per cent	Remarks
Chandrapur (Maharashtra)	12-24 days	30-40	Seeds sown in bags and trays
Chandrapur (Maharashtra)	20 days	25	One year old seed sown in bags
Nagpur (Maharashtra)	15-30 days	22-56	Seeds sown in beds Melghat seeds gave 22%; Yavatmal seeds gave 56% germination
Pakistan (Miani nursery)	40 days	5-8	Seed origin not known
Pakistan (Miani nursery)	36 days	9-12	Above seeds sown in beds.
Pakistan (Khattar nursery)	50 days	30	7 month old seeds from India

It can be seen that fresh seeds gave better germination and germination per cent is higher in beds than in pots or polybags due to better porosity in beds.

(f) Germination Per cent

The germination trials taken out in Andhra Pradesh gave following results.

Table 3.3

(i)	Seed soaked in cold water for 12 hrs	10.0% germination
(ii)	Seed soaked in hot water for 1 hr	9.8% germination
(iii)	Alternate wetting and drying for 48 hrs	12.0% germination
(iv)	Control	6.0% germination

Therefore the seeds put to alternate wetting and drying for 48 hours gave 2 to 3 times more plants as compared to the untreated seeds.

(g) Planting Technique

In Maharashtra the following results were obtained by different treatments:

Table 3.4

	Treatment details	Survival per cent
T1 -	Direct dibbling of seeds	16
T2 -	Planting of 1 year old naked seedlings raised in beds	2
T3 -	Planting of root shoot i.e. stumps of one year old root (2.5 cm)-shoot (22.5 cm)	16
T4 -	Planting of root cutting of finger thickness and 15 to 22.5 cm in length	1
T5 -	Planting of one year old polythene bags plant	73

Therefore polybag plants showed best results. In another trials at Deolapur in Nagpur division, 3 month old naked plants have been found equally good as 1 year old polythene bag plants.

Experiments conducted at Sakoli in Maharashtra indicate that naked planting of seedlings is more encouraging as compared to the direct dibbling of seeds.

In Andhra Pradesh trials were conducted and the method consisted of direct sowing of seeds on the mounds formed in the centre of trenches of size 120 cm x 30 cm x 30 cm. Results of 3 years are given in Table 3.5.

Table 3.5

Year	Survival percentage	Max height growth in cm	Average height growth in cm
1969	69	68	19
1970	67	63	16
1971	82	20	11

(h) Stump Planting

The result of stump planting of tendu in Andhra Pradesh is given in Table 3.6.

Table 3.6

Year	Survival	Max height	Average height
1970	42.5%	2.68 mt.	0.22 mt
1971	56.5%	0.60 mt.	0.15 mt.

In Pakistan, planting of stumps or potted transplants were found better than direct sowing.

(i) Vegetative methods

The vegetative cuttings of tendu has always failed. The pieces of root suckers failed in pots or in beds. The root cuttings were planted in trays in mist chamber after application of harmones but it also failed totally in Chandrapur (1997).

(j) Dying back and Enumeration of Tendu

The tendu forests are heterogeneous in their composition, quality, density and extent. The microclimatic differences produced due to variation in slope, aspect, relief, proximity to water courses, soil and geological formations, cause a very perceptible variation in the vegetation. The tendu is found to occur both in (a) Teak forests and (b) mixed forests, the extent of tendu varies depending upon the micro-climatic variations. The following terms are useful while studying enumeration and dying back of tendu:

(i) **Density:** The absolute density of a tree species is expressed by number of trees (more than 1 m height) per hectare (Trees per 1 ha).

(ii) **Relative Density:** It is expressed as

$$\text{R.D.} = \frac{\text{No. of trees of a species per ha x 100}}{\text{No. of trees of all species per ha.}}$$

(iii) **Basal area:** It is considered as a measure of dominance of the tree species. It is calculated by standard formula (πr^2).

(i) *Absolute Dominance*: This is calculated by multiplying the basal area of an individual species to the number of trees of that species per hectare.

(ii) *Relative dominance or dominance value index (DVI):* This is calculated as follows :

$$\text{Relative Dominance} = \frac{\text{Absolute dominance value of individual species}}{\text{Total of absolute dominance value of all species}} \times 100$$

While it is easy to locate the plants of current years germination, it is not possible to assess the exact age of older plants. Therefore wider age gaps such as 0-1, 1-5, 5-10, 10-20 years of age should be assessed for tendu by the number of scars on the plant stumps because in plants exhibiting dying back the shoot which is produced during the rainy season is ultimately killed in the ensuing dry season. To study roots the roots can be dry around the plants as described by Kolesnikoo (1971).

Root shoot length ratio is determined at various growth stages to ascertain the period of dying back. When this ratio approaches unity it is considered to be the period of all round normal growth and thus is taken to be the period when dying back appeared to cease. Further in practical forestry plants below 15 cm g.b.h. should not be considered as established plants. Under infavourable conditions, particularly where drought is severe, dying back has been reported in *D. melanoxylon* (Troup, 1921). Although it is not possible to provide irrigation to induce natural regeneration in the forest, soil moisture can be maintained at optimum level by manipulation of canopy, cultural operations, mulching and fire protection.

Tendu plants growing under favourable soil conditions as obtained on alluvial soils, continued to maintain lead in shoot length, number of leaves and dry matter production at all ages. The plants growing on stony hill sites were found to be poor in growth rate whereas the plants growing on murumy soil occupied the intermediate position. Shoot length of plants growing on alluvial soil was about two times of those growing on other two soil types. In terms of number of leaves, this difference was 12 times of those growing on dry sites and about 4 times on murumy soil. The root-shoot ratio of tendu seedlings is very high in the first year of planting because the above ground growth is very slow initially. The seedling of 1-5 years age showed a root-shoot ratio of 3.0, at 5-10 years of age this ratio was reduced to 2.6 and ultimately it was 1.2 in 10-20 years.

Though *Diospyros melanoxylon* like Teak and Dhawda (*Anogeissus latifolia*) out numbers other species in plant density, its ultimate progression to maturity is one of the lowest. Thus a high degree of mortality occurs mainly due to biotic factors such as grazing, cutting, leaf plucking and fires. Like other tree species, annuals and perennials, the *Diospyros melanoxylon* in dry deciduous teak forests have been provided with the phenomenon of dying back to ensure its survival and subsequent establishment. The dying back in tendu appears to last upto 10 years of age beyond which normal all-round growth takes place and plants are established (Prasad and Mishra-1980).

Chapter 4
Root Suckers of *D. melanoxylon*

(1) Root Suckers of D. melanoxylon

There are two types of root system in tendu.

(i) Root system of seed germinated plants, and
(ii) Root system of sucker generated plants

In both these classes the root is greyish black and woody. In seed germinated plants the main tap root goes down deep into the soil vertically and could be traced up to a depth of 1.5 metres or more. Generally the root produces lateral branches up to a depth of about 1 metre. These run horizontally as root-suckers. Those at a depth of about 25 cm. run for considerable length. The length of root suckers varies from 2 m to 9 metres and circumference of 5.5 cm is mostly attained for this. The decrease in sucker circumference varies. It may be abrupt or gradual. However the main tap root decreases gradually and remains quite thick.

In sucker generated plants it was expected that plants will be interconnected in the form of a reticulum but it was not so surprisingly and the soil was dotted with thick woody stumps, each with the thickest end free, knotted and showing signs of detachment by decay from the parent branch. The other end after running horizontally up to 2 metres or less goes down deep into the soil. The vertically penetrating suckers invariably formed loops before running horizontally. It may form loops once or twice.

In both the above cases the main stump gave rise to lateral horizontally running suckers which after assuming an uniform

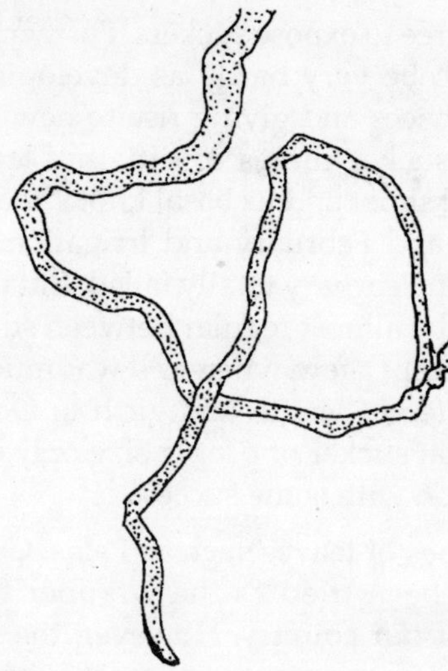

Fig. 7: Loops of Vertically Penetrating Sucker-roots.

thickness went down again into the soil. In their horizontal run these suckers together with the main stump sprouted a few aerial shoots-generally not more than eight. The primary root suckers send forth lateral branches of horizontally running secondary root-suckers which after running a limited distance descend down deep into the soil forming characteristic loops (Figure 7). The secondary suckers after fixing their other end securely in the soil occasionally get detached from the primary branches either by a natural process of decay or by the activities of man. No sooner it happens, the secondary sucker stump sprouts and sends forth lateral branches of tertiary suckers which follows the same pattern of growth and decay. Deep soil fixation is a precondition for any detached sucker stump to sprout. Experiments to sprout cut pieces of root suckers in polybags or in the mist-chamber has completely failed in Chandrapur (1997) and in other many places.

(2) Sucker Plantation

D. melanoxylon is a good coppicer. In the field it forms patchy distribution due to vegetative propagation by root suckers. These root suckers were observed to form a thick network around the parent tree as was revealed by an area of 30 cm x 50 cm excavated around a parent tree to expose suckers. Further more, these suckers were observed to be very hardy as developing and penetrating through rock crevices and giving rise to new aerial shoots. In an experiment the sucker pieces of different length and girth but having one aerial shoot and its basal knot 25 suckers were planted in July, October and February and irrigation was given. The 25 suckers planted in February totally failed. But the sprouting in July was observed to be almost regular between sucker piece length of 20 cm to 30 cm except a few, whereas it was much less and sporadic in suckers of different length and girth in October plantation. It was conceded that sucker of proper size only can be planted only in the rainy season with some success.

Various types of leaves such as Palas, Char, Kumbi, Sal and tendu etc. have been tried as the wrapper for making bidis in different parts of our country. However, the leaves of tendu (*D. melanoxylon*) have been accepted as most suitable by bidi smokers (since they emit the least offensive odour on burning). Tendu is an inferior fuel and fruit trees are seen growing wild in degraded forests and wastelands in abundance in some of the states like Orissa, Madhya Pradesh, Gujarat, Uttar Pradesh and some other states. The young leaves of pre-rain crop (from April to onset of monsoon) are generally plucked as and when they become suitable and are dried in the open sun in the form of bundles. In our country open sun drying is generally adopted at present as it is in-situ, convenient and most economical.

Effect of growth regulators on tendu root suckers

The species of *D. melanoxylon* is generally preferred everywhere for bidi due to its peculiar flavour, flexible and leathery texture, suitable workability when dried and resistance to early decay as compared with the leaves of other species (Desarkar,1963). In an experiment by Negi and Bhandari-1985, the type of treatment given was 5 second alcohlic dip in the basal portion of root suckers of size 15 cm long and 6-8 cm girth. The root-suckers were placed horizontally 2 cm below in the nursery beds of size 1 m x 1 m. Sand and compost in equal proportion was added in nursery beds, root suckers were sown immediately after growth regulators treatment in the last week of May-1982. Judicious watering was given twice a day till break of rains. The result was assessed after 120 days. It was found that:

(a) Optimum concentration of growth regulators 1BA-5000 p.p.m. and NAA 3000 p.p.m. have exhibited significant effect on regeneration of adventitious shoot formation with good flush of leaves followed by IBA 4000 p.p.m treatment.

(b) Concentration of growth regulators i.e. IBA-5000 p.p.m and NAA-3000 p.p.m were quite effective for regeneration of adventitious shoot in tendu root suckers.

The results are clear by the Table 4.1.

Table 4.1: Showing effect of growth regulators on tendu-root suckers

Growth regulators	No. of root suckers sown	Root suckers sprouted	Av. Ht of shoot cm	Av. No. of leaves
IBA-2000 ppm	-	-	-	-
IBA-3000 ppm	10	1	4.0	2.0
IBA-4000 ppm	10	2	5.5	2.5
IBA-5000 ppm	10	5	7.60	4.20
NAA-2000 ppm	-	-	-	-
NAA-3000 ppm	10	5	9.70	7.90
NAA-4000 ppm	10	1	4.5	4.0
NAA-5000 ppm	10	1	4.0	3.0
Control	10	-	-	-

Chapter 5

Use of Tendu Leaves for 'Bidi'

Requirement for 'Bidi' Leaves

The Tendu leaves which are used for 'Bidi' manufacturing should possess the following properties:—

(i) It should be large in size.

(ii) It should be thin and pliable.

(iii) It should be non-pubescent.

(iv) It should be without prominent nerves.

(v) It should have good shining luster when plucked.

(vi) It should have good uniform green colour.

(vii) It should dry up quickly without developing warp, cracks and fracture.

The quality of tendu leaves varies from region to region and from locality to locality depending upon geological formations, soil, topography, climate, weather conditions and extent of biotic interference. The large sized leaves produce more number of 'bidi' wrappers. A leaf can produce 2 to 20 'bidi" wrappers depending upon size, skill of cutting and size of 'bidi' to be made, thin leaves produce less acrid smell and less tar on burning thin leaves facilitate easy rolling due to better pliability. For good rolling of 'bidis' the leaves should be pliable and slightly coarse (like cartridge paper). Pubescent leaves have disagreeable feel and look, and are not liked by smokers. The prominent secondary nerves or veins hinder easy rolling and increases acrid taste. Such veins are scrapped before rolling. The cracks and fractures in a leaf leads to wastage as the cracked portions have to be eliminated while cutting 'bidi' wrappers for rolling. Older stock of leaves make poor quality of 'Bidi'. The bluish green and ashy green colours are considered to be the best, followed by almond colour. Thin leathery texture, pliable, non-pubescent, leaf lamina intact are the desired characteristics for a 'bidi'. Whereas deformed leaf lamina, uneven wrinkled surface, cracks, holes, moulds, pox marks, brown dots on leaves (tipis) and prominent venations are treated as blemishes or defects in Tendu leaves. The 'bidis' made from good quality leaves have good burning quality i. e the 'bidi' once ignited does not extinguish till the end. The time taken to burn one centimeter of 'bidi' and good aroma are the factors which decide the good quality of leaves.

The Tendu leaves suffer from many blemishes or defects which are as under:—

(i) Wrinkles
(ii) Moulds & fungus specks
(iii) Discoloration due to water contamination
(iv) Pox marks or 'Mata'
(v) Cracks, fractures and holes
(vi) Burns, dots in the leaf lamina (Tipis)
(vii) Soiled and dirty leaves
(viii) Blotches due to defective drying
(ix) Immature thin leaves
(x) Over matured thick leaves

The leaves with pox marks, moulds and fungus specks are totally rejected particulary if the leaves are small and no 'bidi' can be made from it. The Tendu leaves are classified as quality I to V depending upon size of leaves and lack of defects in it. Larger the size better the quality, similarly more the defects poorer the quality.

Though the leaves of many other indigenous plants are used for preparing bidis, tendu is the species par excellence which is preferred. The preference is due to certain unrivalled qualities which the tendu leaf possesses, these being:—

(i) Special flavour
(ii) flexible and leathery texture
(iii) easily workability on drying and
(iv) resistance to decay.

Start of Bidi Industry

(As mentioned by A.H. Stein, I.F.S.,D.F.O. North Chanda Division (C.P.) Maharashtra)

Before the year 1915-1916 there was no demand in Chanda division for tendu leaves (*Diospyros melanoxylon*) for making bidis.

In that year the right of collecting tendu leaves in the Government reserved forests of Wairagarh Range were leased out to Pandurang Sao Kale of Sirsi for Rs. 60, this marks the beginning of the bidi industry.

The reasons why a demand for tendu leaves started suddenly in 1915-16 are obscure, but this may possibly be due to the construction in 1913 of the B.N. Railway small gauge line from Nagpur and Handia to Nagbhir and Chanda, which brought the forests of North Chandrapur particularly those of Wairagarh, Gunjewahi, and Brahmpuri Ranges to the notice of the Gondia and Nagpur bidi merchants or to the great war which made cheap foreign cigarettes difficult to obtain their price temporarily prohibitive and thereby induced Indian merchants to commence the manufacture of bidis on a larger scale than before. In the year 1925-26 and 1926-27 the demand of tendu increased suddenly because the bidi manufacturers were making a determined effort to keep the Indian market by the mass production of bidis and by underselling the cheap cigarettes with which foreign manufacturers were flooding the Indian market.

Table 5.1: Production of Tendu leaves and bidis — Chandrapur Division-Maharashtra

Range	Name of place	No. of persons employed for bidi making	Leaves consumed	Out turn of bidi per annum
Chanda	Chanda	60	1,64,25,000	3,28,50,000
Moharli	Bhandak	2	3,65,000	7,30,000
Gunjewahi	Sindewahi	11	18,25,000	36,50,000
Brahmpuri	Brahmpuri	14	32,50,000	65,00,000

These outturns were from government forests and *malguzari* forests. The rate of daily wages or daily earning varied from 6 to 8 annas per day. One anna is equal to six paisa of today.

The revenue realized in Chanda division during that period was given in Table 5.2.

Table 5.2: Revenue realized from Tendu in Chanda division

Year	Total Revenue (Rs.)
1915-1916	50
1916-1917	200
1917-1918	392
....	......
....	......
1924-1925	5485
1925-1926	6605
1926-1927	26,675

The total revenue from 1915-1916 to 1926-1927 was Rs.52,402 only. The tendu was sold either by tender or by public auction in September (since 1923-1924) held at Nagbhid. Uptill 1926-27 the rights to collect tendu leaves were sold by Ranges but in that year the rights were auctioned by sub-Ranges which resulted in very keen competition.

Tendu as a Pulping Material

The *Diospyros melanoxylon* has good pulping property. Sharma, Rai and Guha (Indian Forester-July-1976) compared this spps. with that from Bamboo (*D. strictus*) and formulated in Tables 5.3a, b.

Table 5.3a: Chemical analysis of *Diospyros melanoxylon*

Sl. No.	Particulars	*D. melanoxylon*	Bamboo
1.	Ash content %	0.98	2.23
2.	Cold water solubility	5.30	3.29
3.	Hot water solubility	8.65	6.12
4.	One per cent caustic soda solubility %	13.16	-
5.	Pentosan content, %	16.6	15.06
6.	Lignin content, %	21.1	27.85
7.	Holo-Cellulose, %	79.03	-

Table5.3b: Fibre dimensions

Sr. No.	Particulars	*D. melanoxylon*	Bamboo
1.	Fibre length-mm		
	(a) Maximum	1.94	3.92
	(b) Average	1.14	1.65
	(c) Minimum	0.73	0.70
2.	Fibre diameter-microns		
	(a) Maximum	21	28
	(b) Average	17.5	12
	(c) Minimum	7	7

These figures indicate that *D. melanoxylon* is having short fibre length.

It was further concluded that unbleached pulps suitable for wrapping papers could be manufactured from *D. melanoxylon*.

Chapter 6

Damages to Tendu Leaves

1) Damages to Tendu leaves affecting quality

The Tendu leaves are damaged by the following agencies:—

(i) Hail storm

The hailstorms occurring in April and May inflict heavy damage to juvenile leaves. Heavy hail storm destroys majority of the leaves by tearing and splitting. The hailstorm usually occurs from 15th April to 15th May. It requires following solutions:—

(a) Do the coppicing or pruning operations during last week of March or in the first week of April so that the new flush of leaves come out after the period of hail storm is over.

(b) Coppice the Tendu plants that have been damaged by hail storm occurring in early April.

(ii) Mata, leaf pox or leaf gall

The 'Mata' to Tendu leaves is caused by an insect called *Trioza obsoleta* (Psyllidae- Homeoptera). A rough reddish yellow gall is formed by this insect on the leaves of young plants. Nearly 25% of tendu leaves are destroyed every year by the insect or 'Mata' particularly in the moist years. Occasional rains and cloudy weather invite this damage. If there is no rain and the season is hot, sunny — the 'Mata' attack is non-existing. The juvenile leaves are mostly attacked by 'Mata' making it unfit for bidi wrapping. The control of Mata requires following solutions:—

(a) The coppicing should be done after 'Holi' festival. Before the 'Holi' festival there is likelihood of rains and thunder. The coppicing should be done after 15th of March when the weather becomes hot.

(b) The plants and bushes of Tendu affected by 'Mata' should be cut back. The heavily infested plants must be cutback without fail otherwise it will spread the disease to other plants. The cut back should not be partial - it should be complete.

(c) If the bad weather comes after the coppicing and coppiced shoots get infected by leaf pox in early April then recoppice the same plants to get another flush of new leaves for plucking in May end. After coppicing it requires nearly 45 days to develop the leaves fit for collection.

(d) The Tendu branches affected by 'Mata' should be cut and thrown away.

(iii) Defoliator attack

The defoliators eat the juvenile Tendu leaves. The attack by caterpillars of the defoliators may be localized or may be spread covering 10 to 20 villages. The catterpillars of two types of Butterflies damage the Tendu leaves. The names of adult butterflies are:

(a) Skipper (Family- Hisperidae), and

(b) Baronet (Euthalianaons of family Nymphlyalidae)

The adults although harmless to tendu leaves lay eggs on the leaves and the larvae (caterpillars) on emerging from eggs eat away the leaves. The damages by skippers are wide spread whereas the damage by baronets are confined to localized pockets.

(iv) 'Tipis' or brown dots on the leaves

'Tipis' or 'Tipikis' is a virus disease which result in brown dots on the leaves leading to the lowering of the quality making it unfit for bidi making. 'Tipis' attacks are sporadic and confined to localized pockets. Late coppicing in affected areas solve 'Tipis' attack. The affected Tendu bushes should be cut back and buried or burnt.

(v) Fungus attack

Tender leaves of Tendu are attacked by fungus making the leaves unfit for bidi making. The fungus attack is occasional and is confined to localized pockets. Heavy cloudy weather or continuous showers for two to three days within a fortnight after coppicing increases the chances of fungal attack.

(vi) Leaf Morcha

Due to leaf morcha the leaves get wrinkled, curved and loose colour and thus becoming unfit for bidi making, leaf morcha is possible due to attack by one kind of virus. Leaf morcha attack is sporadic and confined to localized pockets. It is best to cut back and remove these bushes and bury or burn the same to prevent further spread of disease.

(vii) White ants

Tendu plants are occasionally damaged by white ants. Generally the plucked Tendu leaves kept for drying or in storage may be affected by white ants. Use of insecticides will control the white ants.

(viii) Fire

Fire is a good coppicer of Tendu leaves. Fire after coppicing greatly enhances the size and quantity of Tendu leaves. But after the leaves have come up the fire will damage the Tendu crop.

(ix) Grazing

The young Tendu leaves are browsed by goat, sheep, cattle and buffaloes, particularly in drought situations.

(x) Rodents

Rodents do damage to Tendu leaves that are stored in godowns. Rats do lot of damage. The control measures include :-

(a) Use of rat poison.

(b) Clean the floor, sprinkle water mixed with turpentine in ratio 4:1 once or twice, also sprinkle salt and gamaxine in 1:4 ratio then cover the ground with dry leaves or dry saw dust or deoiled rice. Now the ground is ready for storing the Tendu bags. Further sprinkle turpentine in the verandah or on the ground outside the godown. Turpentine treatment along with trapping of rodents very much reduces the rodent incidence inside the godown.

Raised platform can also be used along with the above prescribed treatment if the floor is 'kuchcha' or earthen.

If storing in open then sprinkle water and kerosene in ratio 4:1. Now sprinkle gamaxin and salt in ratio 4:1 over the same- then spread a layer of rejected dry leaves, de–oiled rice bran or 'Palas' leaves (Butea spp.) / Begunia leaves (*Vitex nigandu*) or Neem leaves/(*Azadirachta indica*) or Bamboo mats. Now store the Tendu bags and cover the top with palas, begunia or Neem brush wood.

For controlling white ants in the storage some available insecticide may be used. The insecticide should be mixed with rice husk and sprinkled on the floor and over the bags.

2) Leaf Characteristics

The following table gives the leaf characteristics *D. melanoxylon* for 3 months (P.C. Kotwal-1982)

Table 6.1: Leaf characteristics of *D. melanoxylon*

Characteristics	May	June	July
Area / gm (cm^2)	90.0	89.2	75.0
Chlorophyll-a (mg/g)	1.47	1.55	1.80
Chlorophyll-b (mg/g)	0.83	0.85	1.80
Wt / cm2 (gm)	11.0	11.2	12.9
Total (a+b) (mglg)	2.30	2.40	2.88
Chl. a :b ratio	1.77	1.82	1.66
Calorific value (cals/g)	4279	4318	4375
Ash (%)	6.7	7.0	7.1
Nitrogen (%)	1.09	1.18	1.19
Phosphorous (%)	0.037	0.034	0.040
Potassium (%)	0.16	0.18	0.19
Calcium (%)2.30	2.25	2.35	2.35

From the view point of 'Bidi' making the characteristics of best quality leaves are, area 167.7 cm^2 / leaf, thickness 1.3 mm, healthy leaf of light green colour, fine texture, burning rate 150

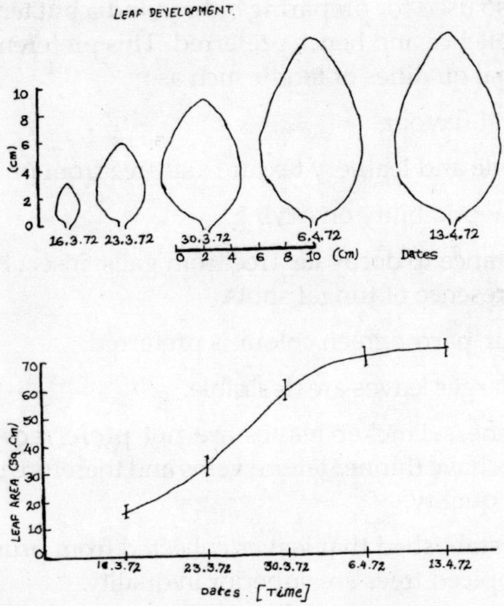

Fig. 8: Showing Leaf Development in Tendu
(a) Weekly Development in Age
(b) Weely Development in Area

seconds / cm of middle portion of 'Bidi' and recovery of 1201 'BIDI' per kilogram of leaves (Patel and Tiwari, 1976). These leaf characters are ideal for 'BIDI' making, although all of these do not coincide together, whereas characters such as leaf area 64.5 cm^2, more thickness1.9 mm, unhealthy dark green colour, coarse texture, slow burning rate 300 sec./cm of middle portion of 'BIDI' or less and recovery of 'BIDI'350 per kilogram of leaves are considered bad for 'BIDI' quality. Most of these characters coincide in April-May when leaves are plucked for the purpose on a large scale and stored dry.

3) Characteristics of good Tendu Leaves

Though the leaves of many other indigenous plants *Diospyros tomentosa*, *Bauhinia racemosa* (A.P. and Goa) *Butea monosperma* (W.B. and T. N.), *Castanopsis indica* (Assam), *Holarrhena antidysentrica* (Goa), *Artocarpus heterophyllus* (Goa), *shorea robusta* (W.B., Assam and Orissa), *Careya arborea* and *Cordia dichtoma* for Chiroot making in Burma are also used for preparing bidis in India but tendu is the species par excellence and hence preferred. This preference is due to the exceptional qualities of tendu such as :

1. Special flavour
2. Flexible and leathery texture and free from pubesence.
3. Easy workability on drying.
4. Resistance to decay i.e. free from galls, insect holes and the presence of fungal spots.
5. Colour-parrot green colour is preferred.
6. Size-larger leaves are desirable.
7. Thickness-Thicker leaves are not preferred. Thinner leaves have thinner lateral veins and therefore the better is the quality.

It is well established that leaves collected from pruned, root suckers and coppiced trees are superior in quality.

4) Quality Classes of Tendu

The tendu leaves have been classified into five quality classes by Sharma and Thakur (1976) as under:

Quality-I : Large sized leaves, thin, smooth textured, weight of leaves being over 160 gm per 100 leaves (i.e. 120 kg per standard bag) . One standard bag = 70,000 leaves.

Quality-II : Fairly large sized leaves, thin, smooth textured, weight of leaves being 91 kg-120 kg per standard bag.

Quality-III : Medium sized leaves (i.e. smaller than those of quality-I and II) fairly thin, fairly smooth textured weight of leaves being 70 to 91 kg per standard bag.

Quality-IV : Small sized leaves, thin to medium in thickness, fairly smooth textured, weight of leaves being 56 to 70 kg per standard bag.

Quality-V : Small sized leaves thin to medium in thickness, smooth or rough textured, weight of leaves being below 56 kg per standard bag.

The soil is an important factor in deciding the quality of tendu leaves therefore the soil of the tendu area after air drying should be analyzed by routine soil analysis (Jackson-1958).

A rating system for estimation of quality of tendu leaves for bidi making has been devised by Patel and Tiwari (1976) wherein seven characters viz. (1) leaf size (2) leaf thickness (3) colour and leaf condition (4) texture and fuel (5) burning quality (6) smoking quality and (7) recovery of usable cuttings per kg. leaves are used. Each class in a character is rated as under in Table 6.2–6.8.

Tables: Rating system for leaf (Patel and Tiwari-1976)

(a) 6.2: Rating for leaf size

Sr. No.	Class intervals sequence Area square centimeter			Rating marks
1	0	to	64	1
2	64	to	77	2
3	77	to	90	3
4	90	to	103	4
5	103	to	116	5
6	116	to	129	6
7	129	to	141	7
8	141	to	154	8
9	154	to	167	9
10	167	to	and above	10

(b) 6.3: Rating for leaf thickness

Sr. No.	Class intervals Thickness (mm)			Rating marks
1	0-11	to	0-13	10
2	0-14	to	0-16	8
3	0-17	to	0-19	6
4	0-20	to	As above	4

(c) 6.4: Rating for colour grade and leaf condition

Sr. No.	Colour grade and leaf condition	Rating marks
1	A	10
2	B	7
3	C	4

(d) 6.5: Rating for texture and feel

Sr. No.	Texture and feel	Rating marks
1	Fine	10
2	Medium	7
3	Coarse	4

(e) 6.6: Rating for Burn Rate

Sr. No.	Class intervals (Time in seconds required to burn one centimeter length of middle portion of bidi)	Rating marks
1	0 to 150	10
2	150 to 180	9
3	181 to 210	8
4	211 to 240	7
5	241 to 270	6
6	271 to 300	5
7	301 to and above	4

(f) 6.7: Rating for smoke

Sr. No.	Characters	Rating marks
1	Strength	25
2	Sweetness	
3	Aroma	
4	General taste of bidi smoke	

(g) 6.8: Rating of recovery of bidis per kg. of leaves

Sr. No.	Class intervals for No. of Bidis			Rating marks
1	301	to	350	7
2	351	to	400	8
3	401	to	450	9
4	451	to	500	10
5	501	to	550	11
6	551	to	600	12
7	601	to	650	13
8	651	to	700	14
9	701	to	750	15
10	751	to	800	16
11	801	to	850	17
12	851	to	900	18
13	901	to	950	19
14	951	to	1000	20
15	1001	to	1050	21
16	1051	to	1100	22
17	1101	to	1150	23
18	1151	to	1200	24
19	1201	to	and above	25

The maximum rating is of 10 points for characters 1 to 5 whereas the same is of 25 for 6 and 7. Total of 7 characters comes out to be 100. Leaves of larger dimension are found on sandy clay to sandy clay loam soils. This may further be associated with low-soil pH, electrical conductivity, organic matter, exchangeable cations and exchange capacity and base saturation. High rain-fall accelerates leaching of base saturation (Buckman & Brandy-1971). Leaf size of tendu becomes small but thick on sandy loam soil. Low exchangeable potash produces thinner leaves nearly of 126 micron (0.1266 mm) thickness.

5) Pathological Problems in Tendu

Dried and processed tendu leaves (*Diospyros melanoxylon* Roxb.) commonly used for preparation of Bidi wrapper, smoking material containing tobacco, is one of the commercially most important minor forest produce of India. India's share in world production of Bidi is 85% which is about 3,00,000 tonnes of Bidi leaves produced annually (Dakwale and Awasthi, 1980). Among the other states Orissa produces about 40,000 tonnes which accounts for nearly 12% of the total annual Tendu leaf production in India (Anonymous, 1976).

The pathological problems were studied in Orissa. For this the gradation of leaves were done into five types taking into account their size and colour arbitrarily by visual differentiation. The graded leaves were categorized into mixed I and II (Leaves of Quality I and Quality II), mixed III (Leaves of quality II and quality III), Quality IV (Leaves of Quality IV) and rejected type Quality V. Tendu leaves are kept in store houses (Phadis) before transportation to market and therefore are liable to further fungal attack which may deteriorate the quality of the leaves (Tuite and Christensen, 1955). The tendu samples of different grades were collected in 1984. Measurement of leaf parameters such as size, weight, area and thickness etc. was done through general methods like measuring by scales, balance, graph papers and screwgauge respectively. Total chlorophyll, chlorophyll-a and chlorophyll-b were estimated following the standard method of Arnon (1949) moisture content percentage of leaves was measured taking 10 leaf discs of 6 mm diameter and drying them at 60°C in oven for 24 hours. Moisture content was calculated following the method of Pixton (1967). Fungi was isolated by moist chamber method (Webster, 1956), leaf washings followed by dilution plote technique (Waksman 1922). Distribution of fungal flora on two surfaces of leaves was measured by leaf imprint method (Potter 1910). Table 6.9 shows various physical parameters such as length, breadth, texture and area of leaves.

The leaves of categories MI and II are good leaves, free from hairs, dirt and holes with few veins, greenish having larger surface area whereas the above characters decreased in other grades resulting coarse bodies, reddish leaves of rejected sample having hairy surface and prominent veins.

It was found that in M-I & II leaf sample contained highest pigment content as well as chlorophyll-a and chlorophyll-b. The chlorophyll pigments were observed to decrease steadily in other

Table 6.9: Leaf Parameters of four qualities of leaves of *Diospyros melanoxylon* Roxb.

Leaf parameters	Grades of leaves			Remarks
	M-I & II	Mixed III	G-IV	G-V
Length in cm	21-26	18-23	17-21	13-16
Breadth in cm	9-12	8-11	8-11	7-9
Weight in gm	15-19	13-16	12-16	10-14
Thickness in mm.	3-4	4-45	4-5	4-5.5
Area in sq. cm	170-190	100-130	90-120	70-80
Colour	Green, Greenish white	Greenish white to greenish gray	Mixed colour	Mixed colour with 90% of leaves brownish red
Other visible characters	Free from hairs, dirt, few veins pliable	Free from hairs, dirt, veins pliable	Slightly having or velvety, slightly prominent veins less pliable.	Coarse hairy or velvety with prominent veins slightly brittle.

samples, the G-V sample containing least amount. Water stress condition and over exposure to sunlight are important factors that cause chlorophyll breakdown in leaves (Haspel-Harvatovie, 1972). The moisture content in G-V sample was low and leaves were found to be quite brittle due to over exposure to sunlight and might be the cause of chlorophyll breakdown. Since sunlight has direct effect on moisture and pigment content, it is essential that care should be taken regarding duration of drying, so that the leaves do not become brittle and discolored due to over exposure.

6) Enumeration of Mycrobial Contamination

Qualitative and quantitative variation of fungal flora of different samples of leaves is represented in Table 6.10 and Table 6.11. Total 24 species of fungi were isolated from the leaves among which various species of *Aspergillus, Penicillium, cladosporium, curvularia* and *fusarium* were common contaminants. Different species of *Aspergillus* were found to be dominant flora in all the samples. Besides *Aspergillus* different species of *Penicillium* were the next important fungus which were present in all the samples except Q-V which might be due to abundance of *A.niger* and *Rhizopus* spp which showed excessive growth rate over others. The relative fungal distribution on surfaces of leaves was studied by leaf imprint method. It showed that the auxial surfaces of leaves supported higher population of fungi than auxial surfaces also reported by Pugh and Buckley (1971) Garg and Sharma (1978) and many others.

Maximum number of fungal species was isolated from M-III and Q-IV. But considering the total population of fungi it was seen that the rejected sample harboured highest fungal population in comparison to others. Fungal population was directly correlated with moisture content of leaves. The moisture content was seen to increase from M-I & II to Q-IV as given in Table 6.12. The moisture content has also a positive correlation with fungal population. The thickness of the leaves may be one of the causes of higher moisture content in Q-IV. It may be inferred that higher moisture content may be the cause of higher fungal population (Pixton, 1967).

Table 6.10: Mycoflora associated with different grades of tendu leaves

Species of fungus	Mixed	Mixed	Quality	Quality
Aspergillus niger	+	+	+	+
A. candidus	+	-	-	+
A. repens	+	+	+	+
A. *flavns*	-	-	+	-
A. clavatus	-	-	+	-
A. fumigatus	-	+	-	+
Pencillium purpurogenum	+	-	+	-
P. chermesinum	+	-	+	-
Cladosporum herbarum	+	+	+	-
C. cladosporiodes	+	+	+	+
Curvalaria lunata	+	-	-	+
C. pallescens	-	-	+	-
Fusarium oxysporum	-	+	+	+
F. *udum*	-	-	-	+
Rhizopus spp.	-	-	-	+

(+) presence of fungus (-) Absence of fungus

Table 6.11: Total fungal population, its percentage abundance on different samples of tendu leaves

Species of fungus	M-I & II		M-lil		Q-IV		Q-V	
	Po/gm	P.A.	Po/gm	P.A	Po/gm	P.A	Po/gm	P.A.
Aspergillus niger	180	46.15	57	13.90	76	10.41	570	47.93
A. Candidus	50	12.82	-	-	-	-	63	5.29
A. Repens	20	5.2	108	26.34	290	39.71	163	13.70
A. Flavns	-	-	-	-	73	10	-	-
A. clavatus	-	-	-	-	75	10.27	-	-
A. Fumigatus	-	-	63	15.36	-	-	43	3.61
Penicillium purpurogenum	9	2.56	-	-	22	3.01	-	-
P. Cherinesinum	22	5.64	-	-	24	3.28	-	-
Cladesporium Gerbarum	23	5.72	32	7.8	55	7.53	-	-
C. cladesporiodes	30	7.69	20	4.87	43	5.75	73	6.13
Curvularia lunata	16	4.1	-	-	-	-	65	5.76
C.pallescens	-	-	-	-	36	4.93	-	-
Fusarinum oxysporum	-	-	20	4.87	27	3.69	20	1.68
F. udum	-	-	-	-	-	-	42	3.53
Rhizopus spp.	-	-	-	-	-	-	110	10.17

Po / gm – fungal population/gm of leaves
P.A – Percentage abundance.

Table 6.12: Percentage of leaf moisture content

No. of observation	Percent moisture content Grades of leaves			
	M-I & II	M-III	Q-IV	Q-V
1	35.82	43.27	44.68	41.45
2	37.75	43.25	44.92	40.62
3	37.20	42.82	45.56	39.95
4	39.90	40.65	46.45	40.31
5	36.90	41.45	46.32	41.15

The Q-IV & M-III leaves had hairy leaf surfaces whereas M-I and M-II were free from hairs. These hairyness of the leaves may be one of the causes of higher fungal population in Q-IV besides the moisture content fountor. Though the Q-V types of leaves had lower moisture content its total fungal population was highest in comparison to others. This may be due to long period of storage during which contamination occurs. Dickinson and Pugh (1974) reported that dried decayed leaves possess maximum mycoflora presumably because these follow the stage of partial decomposition which is known to be characterized by high fungal contamination. *Aspergilhs* were the dominant flora of all the samples. *A. niger* has been reported as a leaf inhabiting fungus by several authors (Kuling 1964, Hagg and Hudson-1966). Meustafa (1971) also reported that *A. niger* was common air inhabiting fungus with a value of 49.9% of total fungal catch of air. *Wahab* (1975) reported *Penicillum* as the second most dominated flora with nearly 18% total fungal catch. *Cladosporium* has been reported as the most common inhabitant of leaf surfaces (Dickinson 1967). *Cladosporium herbarum* which was isolated from all the samples of leaves constituted 7.9% of the total fungal population. It has been reported as most common litter fungi by Subramaniam and Vittal (1980).

A. flavus which is known for secretion of aflatoxin was isolated from Q-IV leaves though in lower percentage of total mycobial population (10%). Since it is a highly dangerous fungus, a further study is needed regarding the presence of aflatoxin in tendu leaves.

All the above results indicate, that M-I and M-II type of leaves could be categorized as best quality in terms of colour, texture, pliability etc with least fungal contamination. The fungal flora was found to be more abundant in Q-IV and Q-V and was the cause for quality deterioration. The various fungi pose serious health hazards to Bidi smokers. The different fungi are commonly associated with storage and are known to have powerful cellulose degrading capacity. Hence prolonged harbouring of various fungi would definitely affect the wall structure and make the leaves brittle. This is the reason why the Q-IV and Q-V type of leaves are not pliable.

The hot and humid climates that prevails in India, particularly during rainy season favours abundant fungal growth which could damage the Tendu leaves under storage. More care is needed for best storage condition for processed tendu leaves.

7) Insects Pests of Tendu

Tendu leaves possess appropriate thickness, texture, colour and pliability, because of which they are preferred over other tree species. The commercial leaves used for bidi making is generally plucked from small and bushy plants because the leaves of bigger size trees are smaller in size and rough in texture. Ram Prasad and Mishra (1980) have estimated a loss of tendu leaves worth Rs. five million from Sagar (M.P) alone due to insects pests. Large quantities of bidi and bidi leaves are exported to various countries earning over Rs. 1 crore worth of foreign exchange (Badola-1989). Such important tree suffers very badly by some insect species which are responsible for colossal loss of revenue and foreign exchange to the country.

(1) Insects attacking the living plants

a. Defoliators

Lepidoptera : *Hypocala rostrata* Fabricions
Miresa albipuncta Hirrich- Schaeffer (Limacodidael)
Euthelia lamdabilis swinhoe
Lamida carbonifera Meyrick
Phycita sp (Pyralidae)

b. Sap-suckers

Homoptera : *Trioza obsoleta* Buckton
Trioza spp (Psyllidae)

c. Root and Shoot borer

Cepidoptera : *Cossus cadambae* moore (Lesidae)

(2) Insects attacking the felled trees and timber

a) Coleoptera : *Ploarderus ferrugineus* Linnaeus
Stromatium barbatum Fabricious (cerambycidae)

The chief pests of tendu are gall forming insect *Trioza obsoleta*, *Trioza* spp, *Hypoclala rostrata* etc.

Gall formers: Tendu suffers very badly by gall forming, insects, the damage of which is locally known as "Mata" or Pox disease. According to Vaishampayan and Bahadur (1986) a mean damage of 50% and 13% to the utilizable leaves occurred in M.P. and Maharashtra and they estimated the losses worth 9 crores and 44 lakhs respectively during 1978 tendu patta season. There are two types of gall farming insects, one close gall farming or pocket gall farming insect known as *T. obsoleta* and other is pit gall forming insects of *Trioza* spp. They are being described as under:

(i) Pocket or Close-gall formers (*Trioza obsoleta*)

Cream white eggs (0.3 mm in length) are laid singly but scattered on the exposed surface of young foliage and buds and are attached to the tissues by means of stalk or process-upon aging, eggs turn yellowish. The incubation period varies from 6-8 days. After hatching young nymphs grow corresponding with the growth of galls. The nymphs are flocculent and exude globules of sugary fluid which attracts ants. The body of active nymph is fringed with filaments. Freshly emerged adults are green in colour but turn greenish to pinkish and at maturity turn bright black. The life span is 4 to 7 days in caged conditions.

The seasonal history of *Trioza obsoleta* is of great interest. Insect completes 5 over lapping generations, two from February to end of April, one during July-August and other two during August to next February at irregular intervals. Symptoms of gall initials on leaves may frequently be seen during July to November.

(ii) Pit-gall forming insects Trioza spp.

Eggs and young nymphs are more or less similar as those of close or pocket gall forming insects. The young nymphs after hatching crawled on the surface for 2-3 days and finally settle at a point mostly on margins of upper surfaces of leaves. As nymphs grow in size, small protuburance on the opposite surface of the leaves are developed. During feeding nymphs exude thick suggery substance in the form of globules which attracts ants. Nymphs moults 5 times before emerging into adults. The nymphal period varies from 15 to 24 days during February and March. The life cycle is completed in 22 to 30 days during March to early April. In April nymphs undergo diapause and reactivate only in next February and thus take about 300 days to complete the life-cycle. The pest completes three generations, two between February to early April and one from April to next February. During diapause period one may spot their presence under magnifying lense.

8) Management of Pest Species

a) Gall forming insects

(i) **Mechanical control**: The population of reactivated nymphs of pit galls and pocket galls can be checked by plucking and destroying the old leaves of April-May in July and that of October-November in late January or early February before leaf fall which varies from locality to locality.

(ii) **Chemical control:** Use of less persistent systemic insecticides may be recommended during late February to March when the pit and pocket gall forming nymphs population remain at their peak. Any insecticide viz. Dimecron, Dimethote, Monocrotophos etc. (25%) can be used.

b) Defoliators

In laboratory conditions Sushil Kumar et al, (1989) tested Monocrotophos 36 EC and Nuvans (100). They found both the insecticides quite effective. They recommended Monocrotophos (0.01%) against the larvae of *Hypocala rostrata* due to its rapid degradability and less persistent nature. In nature Tachanid fly plays a vital role to suppress the pest species. The control of *Miresa albipuncta* and *Labida carbonifone* may be achieved by spraying Monocrotophos or Nuvan (0.02%).

c) Root and shoot borer - (*Cossus cadambae*)

The infected plants should be destroyed. Soil application of Furadon or Phorate 10 G. granules in doses of 5-10 gms. per plant may be recommended during April to July. In tunnels of larvae 10 ml. insecticidal solution of Dichlorophos (0.0125%). Endosulphan (0.05%) may also be injected.

After going through the aforesaid account it can be concluded that gall formers are responsible for the great loss of tender tendu leaves during summer season. In post monsoon flush, defoliator *Hypacala rostrata* cause considerable loss. Other defoliators are of negligible importance. Shoot and root borer has the potential to become major pest in future.

Chapter 7

Coppicing or/and Pruning of Tendu

In order to get good Tendu leaves the small plants and bushes of Tendu are coppiced or pruned. Without coppice or pruning the quality and quantity of leaves can not be increased. The quality and quantity of tendu leaves depend on the Tendu leaves production in the field under prevailing edafic and climatic conditions as well as on the methods employed in various operations i.e. coppicing, plucking, drying, storage, processing, storage and trumpets. The coppicing and pruning of Tendu bushes produce quality leaves of much bigger size. Two methods are employed to improve production and quality of Tendu leaves -

(i) Coppicing
(ii) Pruning and
(iii) pollarding
(iv) Leaves from root suckers.

They are being detailed as under:

(i) Coppicing

Tendu plants when cut at the base produce profuse good quality and bigger size juvenile leaves from the cut back stools and root suckers.

The Tendu plans or bushes shall be cut either —

(a) Flush to the ground
(b) 5 cm above the ground
(c) 10 cm above the ground or
(d) One to two centimeter blow the ground

All the above methods of cutting give excellent results, whereas cut at 20 cm or 30 cm above the ground level gives comparatively poorer results. The growth harmones of Tendu are more at the lowest portion of the stem. Tendu throws out multiple stems after coppicing. These shoots are numbering 2 to 10 or even more. The tendu plants of 5 to 15 cm in girth are coppiced with the help of sharp instrument. It takes nearly 45 days after coppicing to produce fresh leaves fit for plucking. The last date (maximum) of collection of Tendu leaves is 7th June. In view of this no coppicing should be done after 30th April. The monsoon sets in the 3rd week of June and the leaves plucked on 7th June must become dry within 7 days i.e. before 15th June. But drying situation depends on year to year. In some particular year the plucking of leaves is not possible after 20th May as the pre -monsoon showers come in continuously.

While coppicing the stools should not be allowed to be split. The split stools invite disease and switchy shoots–resulting in smaller leaves. It is best to cut flush to the ground with a sharp instrument e.g spade, axe, bill hook or hacker. The coppicing instrument must be sharpened regularly to get a clean cut. The coppicing and surface fire (burning the debris) enhances thinness and increases size of leaves considerably, but fire being inimical to the forest flora and regeneration, it is not advisable to apply fire in practice.

To get more leaves the coppicing may be done two times:

(a) First coppicing in February end or in early March.

(b) Second coppicing in the 3rd or 4th week of April.

The experiment has shown that two coppicings yield large number of leaves. The first coppice giving more number of leaves as compared to the second coppice whereas the second coppice gives lesser leaves but larger leaves as compared to that of first coppice. Larger the size thinner the leaves. While coppicing the weather must be sufficiently hot. The coppicing should not be done if showers are occurring or is likely to occur.

Tendu plants near villages if not cut for 3 to 4 years develop into a strong & hardy sapling called 'Medha' and attain girth up to 25 cm or more. These 'Medhas' when cut back produce several shoots and leaves. Quality of leaves improves in the subsequent years.

The locality factors e.g. Climatic and edaphic factors along with the weather conditions prevailing during February to June greatly influence the quality of tendu leaves produced during a year. The winter rains (before coppicing of plants), pre monsoon shower during March or early April, (after 20 to 30 days after coppicing), followed by scorching sun and hot wind during collection season (15th April to 7th June) favours production of better quality leaves.

The weather affects the production of Tendu leaves as under:

(a) Early monsoon (in late May or early June) shortens the duration of plucking season as a result of which lesser quantity of tendu leaves are collected.

(b) Late monsoon (after 15th June) prolongs plucking period thus giving higher production of leaves.

(c) If the 'Holi' festival is in third week of March the cold season is prolonged in March. This delays coppicing and ultimate production of new leaves, which finally leads to shortening of plucking season.

(d) There should be no rains in the month of March and early April otherwise it will affect the leaves. Within 20 to 30 days of coppicing there should be rains otherwise the leaves will be affected by leaf pox 'Mata' and fungus.

(e) Cloudy weather of 5 to 7 days during unfolding of leaf buds from coppiced shoots leads to 'Mata' attack and wrinkling of leaves.

(f) The showers in late March or early April followed by dry spell leads to more and good Tendu (bigger and thinner leaves) with better texture. These showers also leads to higher number of stems in a plant.

(g) Absence of moisture or late rains and continuous dry spell throughout leads to smaller leaves, curly shaped leaves and poor crop. Due to dry conditions many foliar buds and tender leaves dry up.

(h) The hailstorm (in early April mostly) tears the leaves and there is partial or total loss of crop. Nearly 75% of the crop is lost by heavy hailstorm.

(i) If the hailstorm comes in late April or May, it affects collection and storage of leaves. Many leaves are destroyed before as well as after the collection. Similarly rains destroy the leaves before and after collection. During heavy rains the leaves kept for drying are washed away.

(j) The rains during the maturity period of leaves hastens maturity and leaves become coarse and thick. Therefore the matured leaves should be plucked immediately after heavy showers.

The quality of leaves in the storage in 'phadis' or phadi-centers are affected as below:

(a) Cloudy weather during plucking season impairs drying of leaves in 'phadis' or godowns'. The inner core leaves get discoloured and blotched showing a roasted appearance.

(b) Cyclonic winds and wild storm during plucking season destroys leaves in the 'Phadi' center or drying yard.

The best weather requirement for Tendu leaves are as under:

(i) The winter should be moist.

(ii) There should be rains in the month of January and February. The rains during these two months are brought about by eastern Bay of Bengal monsoon.

(iii) There should be no rains from 1st March to 15th April or so.

(iv) There should be some rains between 15th April and 25th May. The rains in first week of May is highly beneficial.

(v) There should be no rains in last week of May, first week and second week of June.

(vi) The monsoon should not come before 15th June.

(vii) The monsoon may come after 25th June.

(viii) There should be completely dry season in March.

(ix) There should be scorching sun from first April to 15th April.

(x) There should be scorching sun from 15th May to 15th June as far as possible without any rains in between.

a) Time schedule for coppicing

How the climatic conditions i.e. thunder storm, rains, high wind, high temperature etc affect the quality and quantity of Tendu leaves have been discussed in previous pages.

Based on these observations the time schedule for coppicing should be fixed from year to year. It should be kept in mind that just before the 'Holi' festival there is likelihood of thunderstorm and rains therefore the coppicing of Tendu plants should be decided accordingly. The time schedule can be fixed up as under:

(a) Coppicing of Tendu plants just before or after 'Holi' festival i.e. between 20th February to 15th March. Actual period will vary from year to year. The coppicing should not be delayed. The availability of labour should also be kept in mind.

(b) The coppicing should be completed within 15 to 20 days.

(c) If there is need to go for second coppice then it should be practiced between 10 to 20th of April.

(d) Areas affected by pox 'Mata' or hailstorm etc. may require second coppice. In such cases the second coppice shall be made between 20th March and 10th April positively.

(e) The time period of coppicing should be decided by keeping in mind that the development of leaves fit for plucking requires 45 days after coppicing. The number of days required for development of full leaf depends upon the atmospheric temperature. Higher the continuous temperature the lesser will be the number of days. The night temperatures should be more than 20°C.

(f) For getting Tendu leaves from non-forest areas, or agricultural fields the area should be ploughed or coppiced. Light surface fire can also be applied instead of coppicing or just after coppicing. Fire acts as a good coppicer. Fire gives large no. of stems, leaves, in addition to larger and thinner leaves with good lusture. Ploughing and fire should be practiced in the first and second week of March. Application of fire in the forest areas is not possible although it is often mischievously resorted to by the Tendu contractors or purchasers. Remember that applying fire in forest areas is a 'Forest offence' as provided in the Indian Forest Act-1927.

(g) If the area of a 'Tendu Phadi' is large it can be divided into 2 to 4 blocks and coppicing should be made in order-first in block no-1, then in block no. 2, 3 & finally in 4. The same sequence should be adopted while plucking Tendu leaves through the labourers i.e. block 1, 2, 3 and then 4. Further the same block may require traversing 2 to 4 times as matured leaves will go on coming.

b) Development of coppiced shoots

The various factors that decide the development of coppiced shoots are:

(a) Size, pattern and age of underlying root system: The medium aged plants of 5 to 10 years with fully developed superficial roots give large and more number of leaves. Deep underlying roots do not coppice well.

(b) The type of soil decides the nutrient status, aeration and pH value of the soil. Sandy loamy soils, and lateritic soils produce more leaves and bigger size of leaves per stem, whereas black cotton soils have less soil aeration and therefore produce smaller and lesser number of leaves. In clayey and black cotton soils early coppicing is preferable. In quartzite, murumy and lateritic soils the maturity period is less and growth of coppiced shoots is better.

(c) It has already been mentioned that continuous hot temperature is beneficial for the development of tendu leaves. Hot scorching weather promotes larger, thinner and good textured leaves.

ii) Pruning

As far as a particular area is concerned, pruned trees produce larger sized leaves compared to unpruned trees (Kotwal-1978). Therefore the trees in forest are frequently pruned although, not scientifically.

Pruning is one of the important yet often little understood method of securing certain desired results in plants. Some pruning occurs in nature itself. In cultivated plants pruning is done to provide better shade, better screens, better flowering etc. (Everett, 1954). In forestry one of the objectives is to get tall, straight knot free bole which is secured to some extent by pruning. In an experiment the tendu plants of different girths were pruned by Kotwal 1981 and it was found that:

(a) The pruned trees bore quite large sized leaves than unpruned ones. A Marked difference was recorded in leaf characteristics of pruned and unpruned trees as given in Table 7.1.

Table 7.1: Leaf characters of pruned and unpruned tendu plants

(All the twigs upto thickness of 3 cm were pruned off with axe and sickle)

Treatment	Wt/leaf (gm)	Area/leaf cm^2	Area/gm cm^2	Wt/cm^2 (mg)
Pruned	1.210	110.0±8.5	90.9	11.0
Unpruned	0.799	71.4±6.3	89.22	11.2

(b) Pruned trees bore relatively larger and thinner leaves than controls.

(c) It was also found that pruned trees of lower girth classes produced more quantity of leaves (by weight) than controls. But the trend was just reverse in trees of higher girth classes where pruned trees produced less quantity of leaves than controls as is clear from the Table 7.2.

Table 7.2: Leaf biomass (oven dry) of pruned and Unpruned trees

Classes	Plant girth cm	Biomass (pruned) gm	Biomass (unpruned) gm
A	12	0.306	0.222
B	17	0.501	0.451
C	22	0.770	0.654
D	27	1.049	0.829
E	32	1.195	1.204
F	37	1.325	1.337
G	42	1.529	1.677

contd....

Classes	Plant girth cm	Biomass (pruned) gm	Biomass (unpruned) gm
H	47	1.831	2.259
I	52	2.337	2.840
J	57	2.596	3.780
K	62	3.115	5.401

Fall the manufactured foods.

(d) The leaves of pruned trees were more attacked by gall forming insects.

(e) The response to pruning varies mainly with species, age, time, extent of pruning and site quality.

(f) As the root growth depends on adequate supplies of carbohydrates from the crown, the removal of photosynthetic tissue in pruning may reduce its growth during leaf fall.

Tendu leaves suitable for manufacturing of bidis should have large size, thin texture with no tomentosa and practically no prominence of secondary nerves. For this purpose tendu bushes are to be pruned at proper time. The best period of coppicing / pollarding is considered to be the end of February or early March. Leaves are available for plucking 30-60 days after coppicing. Roughly these periods are as given in Table 7.3 (Narukar, 1980).

Table 7.3

Month of coppicing	No. of days required for production of mature leaves useful for bidis
January	55-60
February	40-45
March	30-45

Tendu leaves collection period is very short and starts after 40-45 days of pruning.

iii) Pollarding

In order to obtain a good flush of tender shoots, tendu plants over 15 cm girth are cut at 60 cm to 90 cm height. Pollarding experiments have given good results in Tamil Nadu, Andhra

Pradesh, Maharashtra and U.P. It was found that pollarding produced quantity of leaves and also improved its quality. In Andhra Pradesh the pollarded plot having 201 trees with girth ranging from 15 cm to 70 produced a larger 47 bundles of leaves that were much larger in size and superior in quality as against 27 bundles obtained from unpollarded 230 of trees up 15 cm to 60 cm girth. In Maharashtra it was found that the leaves of the trees pollarded in January and March produced leaves with larger size and better texture. In U.P. annual pollarding is done in Bundelkhand and a profuse crop of leaves is obtained by this operation.

iv) Leaves from root suckers

The leaves produced by root-suckers are much larger in size, softer in texture and are as good in quality as leaves from coppice and pollarded shoots. In Orissa leaves are obtained from 60 cm to 90 cm high root-suckers. In order to get soft and pliable leaves the root suckers are cut back during February and early March. New leaves appear in early April and are generally ready for plucking by the middle of April.

Experiments done in M.P. and Pakistan on trenching around tendu trees to study production of root-suckers were a failure.

Chapter 8

Plucking, Drying and Storage

Plucking, Drying and Storage of Tendu Leaves

Although an effort is made to collect the entire matured leaves in every Tendu season which starts from 15th April and terminates on 7th June or earlier but it is not possible to collect 100% leaves. Nearly 30 to 40 per cent tendu leaves have to be left in the field either due to smaller collection period available or due to lesser availability of forest labour. After inspection of the area a planning chart needs to be prepared so that all the leaves may be collected in a season. In this planning the time of coppicing should not be included as the time period of coppice is dependent on climatic conditions in the month of February, March and April. The number of pluckers and distance they will be covering, drinking water facility, medical facilities to labour (plucking season is prone to heat stroke, malaria, and flue), child care and care of female labours, shades for resting, cycle facility to phadi munshi etc have to be taken into account.

One average plucker usually collects around 6000 to 7000 leaves per day or about 1.8 lakh leaves during a month. Nearly 7 to 10 man days are required to coppice an area capable of producing 1.5 lakh leaves, depending on the incidence of Tendu bushes in the concerned patch (i.e. 7 to 10 man days for coppicing per plucker) – (Sarangi - 1990).

The Tendu leaves collection has been nationalized since- 1969. Any Tendu plant producing Tendu leaves for bidi wrapping either occurring in Government forests or in the private farm or field is a Government property. There are two methods of tendu leaves trade:

(i) Through the purchaser in which case the Tendu units (consisting of 8 to 12 phadi centers) are auctioned. After auction it becomes the responsibility of the purchaser to coppice the area and pluck the leaves as per rules and norms fixed in the sale condition.

(ii) Departmental collection in which it is the responsibility of the forest department to coppice the area, arrange labours for plucking, drying, storing and finally auctioning the leaves. Therefore coppicing is performed in supervision of the Range Forest Officer, Forester and Forest Guards according to the guidelines obtained from the higher authorities.

In both the above methods correct method of coppicing and effective time schedule should be employed. A party of coppicer may have 10 to 20 labourers. No area should be left for coppicing otherwise it will give decreased revenue.

The sharpened axe is generally the instrument for coppicing. The number of axe-man will depend upon area of coppicing, terrain, soil condition, incidence of stones and pebbles, age and size of Tendu stems, mixture of Tendu plants with other bushes and plants (thorny bushes of other species are hard to deal with), direction of labour movement, weather conditions (very hot weather will decrease the efficiency), stamina of labours and create other inhospitable conditions.

The area of coppicing can be divided into 3 to 4 blocks and each block can be handled serially in a week or so. Further each block can be divided into 3 to 4 sections, one section should be coppiced in one day positively. A coppiced area should be immediately checked to coppice the left over plants. The potential for production of leaves should be determined based on the last three years experience.

a) Pruning

In pruning operation the branches of the crown of the tree or plant is sharply cut or the stem is pollarded to produce juvenile leaves.

The pruning operation may be carried out for older plants of Tendu of 8 to 12 years of age, in the month of late February or early

March. The pruned plants or trees will produce new flush of leaves which will be bigger as compared to previous naturally grown up leaves from unpruned plants or trees. The leaves obtained from pruned saplings are inferior in quality than the leaves obtained from the coppiced plants. Only in case of crisis and shortage of leaves the leaves from pruned saplings should be collected for 'bidi' making. Leaves on the pruned trees take 55 to 60 days to mature.

b) Root Suckers of Tendu

Tendu shoots can also be obtained from root suckers. The superficial roots of Tendu when injured by sharp instrument give out root suckers. The superficial roots are cut in concentric circles at certain spacements by keeping the original tree at the center to induce growth of root suckers in the coming monsoon. Only the secondary or peripheral roots should be injured or cut. After a period mostly in the coming monsoon the new shoots come out from cut roots or root suckers. These shoots produce larger leaves on the first occasion but when coppiced further in subsequent years produce better and very good quality leaves. The root suckers should be treated as independent plants of Tendu and should be coppiced every year. After coppicing for more than two years the coppice shoots provide very good quality leaves.

c) Plucking of Tendu Leaves

The Tendu leaf collection season starts in Aprii and terminates in the last of May or early June. The actual period available depends upon matured leaves availability, climatic conditions etc. so that drying of leaves is possible. Generally the period of collection prevails from 15th April to 7th June. Before plucking it should be ensured that facilities for drying of leaves and shade or godown for storage is available.

The tender tendu leaves are dark carmel or dark red or a mixture of both but with the passage of time as the atmospheric temperature rises the leaves become greener and greener. A fully developed green leaf is a sign of maturity. In older and higher plants fully matured or over matured leaves are green, thick, rough and leathery. Such leaves are not fit for bidi wrapping. Whereas semi-matured slightly soft and greenish leaves are fit for bidi making.

The immature leaves lack total greenness, they are thinner and soft, with dorsal surface shining, leaves pendulous (i.e. hang down), ventral surface villous hairy or highly pubescent, leaf when torn has a sharp or uniform edge, if the leaf lamina is pressed between finger nails the leaf gets cut, tip of the immature leaf on drying gets curled and wrinkled, petiole is tender, dried immature leaf on moistening and crumpling by hand breaks into bits.

On the contrary the entire or semi-matured leaves do not have reddish tinge but are deep green to olive green, slightly thick and coriacious i.e. leathery to touch, dorsal surface is rather dull, leaves do not hang down - with the maturity young hanging leaves become firm, ventral surface are glabarous (without hairs) or slightly pubescent, if the leaf is torn from apex the torn end has a slightly irregular dentate edge, if the leaf lamina is pressed between finger nails and seen through the light, minute veins are seen and leaf does not get cut, the tip of the leaf does not curl on cure drying petiole, firm-not straight but slightly curved.

It is to be borne in mind that over matured leaves are unfit for bidi making. The over matured leaves occur on older saplings and on trees. The distinguishing features of over matured leaves are:

The colour is dark green, veins on the ventral surface is very conspicuous, texture of the leaf is rough and thick, petiole is thick and straight, the leaf surface is uneven and depression is marked along veins.

d) Season and Time of Plucking

The season of Tendu leaves plucking starts from 15th April and remains up to 7th June or so. The actual period is dependent upon weather conditions. The rains create hindrance in plucking and drying of leaves.

The time of plucking starts with sunrise and remains till the sunset. In hot afternoons the plucking is very exhausting due to high atmospheric temperature. The labourers therefore should be employed from 6 hr. to 13 hrs. in the morning and from 15 hr. to 19 hrs. in the evening.

e) Organizing Tendu Leaves Plucking Through Labours

The women, children, young children, old men and women are mostly employed for plucking. The leaves are plucked by hand directly from coppiced shoots by the employed labours. First of all the good quality leaves are plucked by leaving smaller leaves to mature. The leaves are thus plucked in three installments, about 50% of the leaves are plucked in first installment, 30% in the second installment after 10-15 days and 20% in the third installment. Thus a plant or bush is traversed three time as it is not possible to collect all the leaves at one time. The collected leaves are carried home in head loads. An average plucker can collect nearly 20 kg for one head load whereas a strong person can carry up to 25 to 30 kg by head load. Nearly 10,000 to 12,000 leaves can be collected in 4 to 5 hours. An average load contains 6000 to 8000 leaves. The Tendu collection provides income to the poor villagers during summer season when no other employment is available. The labourers have to travel on foot in the jungles for collection and transport by head loads. The income per day of a labour is generally higher as compared to the other works in the same village. Generally females earn more than the males.

f) Phadi Munshi

The collection center of tendu leaves is known as a 'Phadi' in local parlance. Each 'Phadi' is supervised by a phadi munshi or 'agent' who is the employee of the tendu unit purchaser or contractor. He organizes the labour, leaves collected per day, leaves stored in the local godown and the leaves despatched for central godown. The agent or the phadi munshi maintains the necessary records and registers for the control of tendu leaves and makes payment to the labours. In a departmental collection these works are allotted to the Forester & Forest Guards. No payment is made for under matured and over matured leaves as they are unfit for 'bidi' making.

g) Tendu Units

A Forest division is divided into 5 to 7 Forest Ranges. The Range Forest officer is the incharge of a Range. A Forest Ranger or R.F.O has 3 to 6 Tendu units spread over the area of the range. A 'Unit' has 8 to 12 collection centers or 'Phadis'.

h) Leaf Bundles

The collected leaves are tied in bundles. A bundle contains 70 leaves, 5 less or more i.e. 70 ± 5 leaves. The leaves are put in a way that 35 leaves face one side and the other 35 leaves i.e.50% leaves face the other side. The leaf bundle is tied with a thread or grass. The bundle is known by different names eg. 'Guddi', 'Puda'or 'Kerri'.

i) Drying of Tendu Leaves

The leaf bundles or 'Pudas' are placed on the sandy ground for drying. While drying these bundles are turned upside down so that only top and lower leaves are exposed to the sun alternately. The middle 70-2=68 leaves get dried without direct exposure to the sun on the surface. The leaf bundles of 70 each are laid on the ground 15 cm x 15 cm apart in 'Chatta' or 'Chokdi' in group of 200 pudas i.e. 10 lines x 20 lines or in any other convenient fashion depending upon the local practice in vogue. To facilitate drying by sun the bundles are so placed on the ground that long axis of each one of them always runs East-West. By this arrangement the solar radiation coming from southern direction penetrates deep into each leaf bundle and dehydrates the leaves rapidly and uniformly. The grouping of leaves into 'Chatta' or 'Chokdi' helps in counting the bundles. The inspecting officer or checking officer finds it easy to count the each days collection.

As the days advance the colour of the drying leaves go on becoming darker and darker and it is very easy to distinguish last seventh days collection from last six days collection and right up to todays collection.

The 'Phadi' center or the drying yard should have following requirements:

(i) Drying place should be a level or slightly sloping land having no shade at all.

(ii) The place should be dry and sandy ground should be preferred.

(iii) The ground should be free from insects and white ants and it should be located within one to two kilometres from the village.

(iv) The place should have capacity for placing 7 lakh to 14 lakh 'Pudas' or bundles because each day nearly 1 to 2 lakh bundles may be collected and each days collection requires 6 to 7 days drying.

(v) The place should not be a grazing ground for cattle, cattle route, playing grounds for children, timber or fuel wood depot etc.

(vi) The place should not be vulnerable to floods if the place is near a river.

The place should be divided into seven parts for collecting each day's bundle.

Proper drying ensures good quality leaves. Well dried leaves do not succumb to fungus or insect damage, drying includes turning process. One turning per day should be made without fail.

The bundles affected by rainwater are shaken off by beating the same lightly once or twice on the hand and putting back the same on the ground. If the ground has become wet, the bundles are carried and spread again on nearby dry ground. The outer two leaves get spoilt due to exposure to sun, they act as cover leaves. The loose bundles are tied with the strip of palm or Sabai grass. A leaf bundle is completely dry when all the petioles become white and brittle. If the petiole instead of breaking gets bent on application of pressure the bundle is not dry and these should be allowed to dry in the drying yard for one or two days more.

The green leaves loose weight on drying from 55 per cent to 65 per cent of green weight of the leaves. The lack of turning of bundles in time leads to unequal drying, loss of colour and loss of required aroma during smoking of bidi. On the last two days of drying the bundles are arranged. Vertically in close concentric rings kept standing on petioles. This makes hot air to dry the petioles and mid rib of leaf lamina. This method is called 'Dhera' drying.

j) The Bag, Trade Bag, Hessian Bag and Standard Bag

The dried up tendu leaf bundles are moistened by sprinkling water over it before putting in the cloth bag or Hessian bag which are also called trade bags. A bag, trade-bag or (Hessian bag) is sufficiently big to contain 400 to 700 bundles depending upon the size of leaves. The bundles should not crack while putting in the bag. The bag is stitched with sutli. Under-dried bundles should not be kept in the bag. Under drying creates blotchy colour. In worst cases the core leaves get discoloured and fungus affected. In such cases the bag is reopened and the leaves are further dried. If necessary rebundling is also done.

A standard bag means 1000 bundles of 70 each. One standard bag contains 70,000 leaves of Tendu.

k) Storage of Tendu Bags

The bagged Tendu bundles are then stored in the godown or sheds. The godown ground should be cleaned and made termite proof. The godown should be rain proof and moisture proof. A tarpauline on the ground and also on the top may be used. The common salt or insecticide may be sprinkled on the ground. The godowns can be 'Kachcha' or 'Puckka'. In any case some damages are likely to occur in the transport and storage process.

Chapter 9
Processing, Packing and Sale

Processing, Packing and Sale of Tendu Leaves

After drying and softening the bundles or 'Pudas' are put into trade bags which are generally made of hessian cloth i.e jute fiber. Each trade bag after being filled weighs 50 to 60 kg in which nearly 500 to 600 bundles are packed. It is necessary to soften the bundles by water sprinkling on heaped bundles before packing otherwise the bundles will crack. The moist bundle heaps should be kept covered with tarpaulins or bamboo mats for about 5 to 6 hours till dry bundles get moist and soft or pliable enough to be handled without cracking. Adding too much of water during 'softening' process makes the leaves liable to fungus attack for future storage. The correct amount of 'softening' is checked by feel of the hand of the bundles. The moistened bundles are packed tightly into the gunny bags. Hand pressure and rolling of bags is used to fill maximum number of bundles. Tight packing is also necessary to avoid cracking and breakage of leaves during transit. Nearly 30 rejected leaves are put at the bottom of the bag and same number of leaves are put at the top before sewing the gunny bags. After filling the bag weighs 50+10 kg depending upon the size of the leaves. The size of a gunny bag is 112 cm x 270 cm. The gunny bag has 11 porters x 9 sorts per sq. inch and these bags can hardly be used for any other purpose except for filling the Tendu bundles. The trade bags (gunny bags) are transported to far off situated central godowns in taluka or district places by means of trucks. Before transport each bag is given:

Forest Division Name ....

Range Name ....

Name of Purchaser ....

Unit No. .... Village Name (Unit name) ....

Phadi No. ....

Total no. of 'Pudas' ....

Name of Phadi agent ....

Year of Collection ....

Name of Godown ....

These descriptions are written by means of water soluble colours on the hessian bags.

The transported bags are placed inside 'Kaccha' or 'pakka' godowns hurriedly before the onset of monsoon rains in June. During transport the bags should not become wet therefore necessary precautions needs to be taken for this. During the storage also the bags and bundles should be checked for mildews or fungus attack. The fire extinguishers and water filled buckets should be placed near the godowns to deal with the fire hazard.

Before selling the bundles may be rearranged quality-wise in the bags by throwing away the defective bundles or the Tendu leaves can be sold as it is. The good leaves are also likely to be discarded and thrown away due to the carelessness of sorting labourers, therefore the waste leaves should also be inspected and utilizable leaves should be recollected. Similarly the processed bundles may contain useless, lower quality leaves or less leaves therefore it is necessary to check processed bundles before packing into gunny bags. A group of female labour together with male labour and some children constitute a group for sorting and bagging.

Sorting and grading of leaves is necessary according to market requirements. The bundles are graded into the grades I, II, III, IV, and V. These are placed in separate bags grade–wise. The purchasers or whole sellers buy different grade Tendu leaf bags and sell the same under different trade names. The dealers in Tendu leaves are very rich persons and cheating by one or the other way means loss of business for the future. To ensure correct bagging a token is inserted in the bag which bears the name and signature of bagging-labourer responsible for bundling and bagging with correct grades of leaves. Tendu leaf purchasers are very selective and complain badly in case of defective grading and inclusion of inferior grade leaves or useless leaves in processed bundles and bags. Higher-grade leaves should be handled very carefully giving no chance for complaint. Particularly in case of export of quality leaves to foreign countries one has to be extremely careful. Strict supervision and control is therefore very necessary. On receipt of orders the bidi leaf contractors open up central phadi houses and take up processing, grading, bundling, packing and bagging according to designated specifications under their personal supervision. This is called central processing, and binding. The Tendu bags should be checked regularly and put to sun twice or thrice in a year to ward-off fungus and insect attack. The damaged bundles should be replaced, without playing any gimmick.

Chapter 10
Trade of Tendu Leaves

1) Trade of Tendu leaves

The tendu leaves are utilized by 'bidi' factories for manufacturing 'bidis'. The labourers are issued one to five kg of Tendu leaves along with tobacco ('Jarda'-local name) for rolling 'bidis' at their residence. The number of 'bidis' to be made from one kg of 'bidi' leaves depends on the size and thinness (quality) of tendu leaves. Many 'bidi' factories engage workers to role 'bidis' in the factory itself under their supervision instead of allowing the workers to roll 'bidis' in their residence. The workers moisten the 'bidi' leaves by sprinkling water and covering the same with a gunny bag. Once the leaves become soft, useless leaves are discarded. Only good leaves are cut by expert hands. The size and dimension of leaf cuttings vary according to length and size 'bidis' to be manufactured.

The appox. estimated production of tendu leaves in different states are :

Andhra Pradesh	-	40,000 tons
Bihar	-	25,000 tons
Gujarat	-	5,000 tons
Madhya Pradesh	-	1,66,000 tons
Maharashtra	-	50,000 tons
Orissa	-	50,000 tons
Uttar Pradesh	-	10,000 tons
Rajasthan	-	6,500 tons
Tamil Nadu	-	3,000 tons
West Bengal	-	500 tons
Karnataka	-	500 tons
Kerala	-	500 tons
Total		3,57,000 tons

The production of tendu leaves each year changes and is dependent upon climatic conditions during tendu season.

Tendu has assumed a great economic significance mainly due to the industrial use of its leaves for 'bidi' manufacture. At present leaves are collected from more or less degraded vegetation which is being exploited time and again without any tending and other operations to improve production. Scant attention has been paid towards research on various aspects of leaf production barring some coppicing, pollarding and plantation techniques trials in two or three states.

The tendu trade is a nationalized trade since 1969 and tendu plants and leaves wherever growing whether on the Government land or in the agricultural fields are Government property and purchaser of the Tendu units encompassing the Government or private area is entitled to collect the Tendu leaves both from Government lands as well as from the private agricultural fields. The state Government conducts 3 to 4 auctions of Tendu leaves through the Forest Department of the states in the months of:

November	-	1st auction
December	-	2nd auction (for unsold units in Nov.)
January	-	3rd auction (for unsold units in Dec.)
February	-	4th auction (for unsold units in Jan.)
March	-	5th auction (for unsold units in Feb.)

If still some units remain unsold due to remoteness of the area or due to higher upset prices fixed by the forest department then those Tendu units are ordered to be collected departmentally. In the departmental collection the Tendu leaves are collected and stored in the godowns of the forest department then they are auctioned or sold by negotiation (if the repeated auction fails) to the traders/ purchasers of the Tendu leaves.

The sale of Tendu leaves fetches enormous revenue to the state forest department every year. In some of the states Tendu leaves are sold through the Forest Development Corporations of the states e.g. in Orissa. In both the cases auctions are held as per sale conditions approved by the Government.

The State–wise trading Centers are:-

Andhra Pradesh	-	Nizamabad, Karimnagar, Mahboobnagar, Mancherial, Warangal, Cuddapah, Kurnool, Anantpur etc.
Bihar	-	Chakradharpur, Jhabua, Muzaffarnagar, Bhagalpur, Bihar Sarif
Gujarat	-	Bhavanagar, Ahmedabad, Baroda, Godhra, Mehsana, Borsad, Nadiad, Surat.
Karnataka	-	Mysore, Mangalore, Belgaum, Nipani, Bangalore etc.
Kerala	-	Cochin, Cannanore, Kazhikod, Trivendrum, Quilon, Palghat.
Madhya Pradesh	-	Jabalpur, Raipur, Bhopal, Indore, Sagar
Maharashtra	-	Nagpur, Kamptee, Gondia, Bhandara, Chandrapur, Gadchiroli, Nasik, Sinnar, Poona, Mumbai, Kolhapur, Satara, Sangamner, Tumsar, Wardha, Amravati, Nanded, Aurangabad.
Tamil Nadu	-	Madras, Salem, Tirunelveli, Trichirapalli.
Orissa	-	Sambalpur, Cuttuck.
Uttar Pradesh	-	Jhansi, Mirzapur, Varansi, Jaunpur, Allahabad.
West Bengal	-	Murshidabad, Dhulian, Calcutta, Raniganj, Siliguri, Darjeeling.

There are wide variation in the quality of leaves in different states and different regions within a state. Leaves from the northern part of Orissa are much better in quality than those from southern part, consequently fetching a price five times that of the later.

Similarly quality leaves are produced in Bilaspur district of Madhya Pradesh which are very large in size, papery in texture, non-brittle with inconspicuous veins. There may be many more areas capable of producing high class leaves which need to be identified. The need for organizing systematic research and various aspects of cultivation of tendu plant is of great importance.

The glossy tendu leaves are preferred in the trade whereas it is very in difficult to sell pubescent 'bidi' leaves. Poor quality 'bidi' manufacturers take pubescent Tendu leave. The good quality leaves are thinner and produce less tar or ash in burning. The good leaves produce good aroma while smoking. But ultimately 'bidi' manufacturers have their own choice in respect of Tendu leaves. The choice depends on:–

(1) **State and district:** which is supposed to be good, lucky and auspicious for the manufacturer.

(2) **Unit no:** which is supposed to be lucky number based on the numerology for the manufacturer.

(3) Taste and aroma of the Tendu leaves.

(4) Convenience in business handling and easiness of handling the trade.

The first year Tendu leaves fetch maximum price and are sold readily whereas second and subsequent years stored leaves fetch lesser prices which go on decreasing as the time elapses.

Tendu leaves have assumed great economic importance owing to the flourishing 'bidi' industry, which was first started in India during 1887 and has now 2171 factories which manufacture about 140 crore 'bidis' per day. About 7-8 lakh people are employed in leaf-collection and about 85% of total world production of 'bidi' is from India. More than 3.5 lakh tonnes of 'bidi' leaves are collected every years. India has monopolized in 'bidi' production and its export trade. During 1972-73, India exported 'bidi' leaves and 'bidis' valued at Rs. 1.10 crores and Rs. 0.14 crore respectively. It is estimated that about 30 lakh rural poors are employed in 'bidi' making. In the year 1981-82, the Government earned over 6.3 crore rupees as foreign exchange by exporting 52593 quintals of tendu leaves (Dharetal 1989, Tiwari- 1981). Tendu leaves are one of the important components in 'bidi' industry where in almost no expenditure is involved in making infrastructure.

Besides internal consumption large quantities of 'bidi' leaves are exported to various countries like Afghanistan. Bangladesh, Sri Lanka, Nepal, Singapore, Canada, U.K., U.S.A. and several African countries.

The Tendu leaves and bidis are exported to Ceylon, Bangladesh, Nepal and other Countries from India. Nearly 10,000 tonnes of Tendu/ (bidis) are exported annually. The exporting centers are Amritsar, Delhi, Mumbai, Calcutta, Dhulian, Raxaul etc.

'Bidis' are injurious to health and bring many diseases but the main points in consideration are:

(i) Tendu leaves give large-scale employment to poor rural people and the Adiwasis living together in the forests.

(ii) Tendu leaves fetch handsome revenue to the various state Governments.

(iii) 'Bidis' provide foreign exchange to the Government of India and to the Indians.

Every effort should be made to collect maximum Tendu leaves from the Indian forests wherever it is occurring. The plantation and natural regeneration of Tendu should be encouraged in Tendu bearing areas as well as in non-tendu bearing areas.

2) Tendu Trade-Economic Significance

The economic significance of tendu leaves lies in the fact that it is produced and consumed by the rural poor. Besides internal consumption large quantities of bidis and bidi leaves are exported to various countries- earning over 62.729 million Rupees of foreign exchange by exporting 52.593 quintals (1981-82). Two *Diospyros* species viz. *melanoxylon* and *tomentosa* are much favoured and extensively used in bidi manufacture throughout India as wrappers. Both the aforesaid species of *Diospyros* congregate in cool, sheltered valleys and elsewhere prefer loamy soil with dominant clay fraction. The trees develop best on soils derived from metamorphic rocks. It is a conspicuous constituent species found in all the major type of *teak, sal* and mixed forests. It can be grown by seeds, root suckers and coppicing. Observations show that naked planting of 3 months old nursery seedlings give good results. A properly grown seedling can start yielding leaves useful for bidi at the age of 5 years. Root suckers may turn into trees but their rate of growth is slow (Narukar-1980).

As present India is the biggest bidi producer in the world and also the biggest consumer of this commodity; about 88,000 tonnes of bidis are being consumed in India annually. While bidi consumption within the country is enormous India has the monopoly in export of bidi leaves and bidis. During 1972-73 foreign exchange of worth Rs. 1.24 crore was earned from export of bidi leaves and bidis. At present the whole yield of tendu leaves is obtained from the scattered natural growth which involves a high cost of collection.

India exporting bidis to Canada, America, Belgium, Netherlands, Switzerland, Australia, Sri Lanka, Nepal, Bangladesh and Gulf countries.

3) Employment in Tendu

It is observed that the employment generation capability of tendu leaves involving collection, its drying, packaging and transportation are for a period of 6 to 8 weeks (seasonal) and 300 days (regular) in manufacturing of bidis in a year. On the average manpower requirement for one tonne of dried bidi leaves is as below (Pant 1984) :-

(a) Plucking and bundling	-	95 mandays
(b) Drying	-	3 mandays
(c) Packing (after moistening)	-	5 mandays
(d) Checking staff	-	4 mandays
		107 mandays

If the estimated production is 0.4 million tonnes then the employment potential in collection will be:-

= 0.4 x 107 million man days

= 42.80 million man days

In tendu industry about 50-70 per cent of the total selling price go as wages. Wages being the major factor even a minor change in respect of wage rate affects the price structure of the bidis. It is a source of additional income to agricultural workers during their off season. No. of man days required to manufacture bidis from 1 tonne of tendu leaves (Pant-1984) is:

= 180 man days

Employment potential in manufacturing of bidis from 0.4 million tonnes of tendu leaves:

= 0.4 x 180 million man days
= 72.00 million man days

References

Athya, C.D.(1980), A. Study of regeneration of *Diospyros melanoxylon* with special reference to sucker growth and leaf quantity.

Awasthi, Ajai Kumar, *Diospyros melanoxylon* in its forest community-Journal of *Tropical Forestry*-Oct-Dec.1987

Anon (1976-82), Bidi Bulletins-All India Bidi Industry Federation, Bombay,

Anonymous (1976), Our kendu Leaves. A booklet published by Govt. of Orissa Forest Department.

Brandis, D. (1874), The Forest Flora of North-West and Central India-London.

Cooke, T. (1901-1908), The Flora of the Presidency of Bombay, London.

C.S.I.R. (1952), The wealth of India-Raw Materials - Vol. III; D-E, p.-81.

Desarkar, B.K.(1963), A note on Bidi leaf (*Diospyros melanoxylon*). *The Indian Forester* Vol. 89 (I);39.

Dictionary of the Economic Products of India- Vol.- III -p-147-149

Dakwale, R. N. and A. K. Awasti (1980), Bidi Plant–a green coin of forests of Indian Peninsula -*Scientific Reporter* 17 (2): 117-119

Ghosh, R.C., N.K. Mathur and R. P. Singh (1976), *Diospyros melanoxylon*, its problem of cultivation- The *Indian Forester*-June-1976.

Hooker, J.D. 1882 (1872-1897), The Flora of British India.

Indian Forester-May-1904, The Genus *Diospyros* in Ceylon-p-218

Indian Forester-August-1976, All India Seminar on BIDI-May-1976-at Jabalpur, M.P.

Kotwal, P.C., Pruning and leaf production in *Diospyros melanoxylon* - Roxb. *Annals of Arid zone* -20 (2), 117-121 June-1981

Kotwal, P.C. (1973), Ecology of *D. melanoxylon*- Roxb with special reference to its productivity. Ph. D. Thesis - Vikram University Ujjain.

Mishra, R (1968), Ecology work book. Oxford and I.B.H. Publ. Co. New Delhi, Bombay.

Mishra, K, (Miss), Dr (Mrs) A. Mishra and G. S. Padhi, Studies on quality gradation and mycobial contamination in processed kendu (*Diospyros melanoxylon* - Roxb) leaves. The *Indian Forester,* August 1986 - p. 687.

Kumar, Sushil, Prasad Lalji and Khan, H. R. (1989), Seminar on forest protection, 29-30 June 1989 at F.R.I Dehradun - pp.1-6.

Namdeo, R. K., Arun Prasad, G. N. Kharakwal, A. K. Bhowmik, P.K. Khatri, S. S. Bisen and N. G. Totey (1991), Preliminary Study on Physico - Chemical properties of soil as related to Tendu (*Diospyros melanoxylon*) leaf quality for bidi making - *Indian. J.MFP.* Vol.-I 42-51.

Negi K.S. and Bhandari A.S., A note on Regeneration of shoot in Tendu (*D. melanoxylon*- Roxb.) Root suckers by growth regulators. The *Indian Forester* - June 1985.

Prasad, R. & G. P. Mishra (1980), Ecological status of *D. melanoxylon* in dry deciduous teak forests of Sagar (M.P). The *Indian Forester* 106 (I), 41-25.

Rathode, J. S. - *D. melanoxylon*-Roxb, Quantitative Morphology of leaf forms- The *Indian Forester* - Dec. 1976.

Rathode, J. S (1968), The morphology and Ecology of 'Bidi' plant - *D. melanoxylon* (Roxb.) Ph.D. Thesis, University of Sagar, M. P. India

Ram Prasad and G. P. Mishra (Jan. 1980), Ecological status of *Diospyros melanoxylon* in dry deciduous Teak forests of Sagar (M.P). The *Indian Forester*.

Stein . A. H. -IFS - D.F.O. North Chanda Division C. P., Brief, Note on the Bidi Industry in North Chanda Division- C.P. The *Indian Forester*. April. 1928.

Sushil Kumar, Insects Pests of Tendu (*Diospyros melanoxylon*) and their management, *Indian J.MFP.* Vol. I: 92-98, 1991.

Troup R. S. (1921), The Silviculture of Indian Trees. Vol. II

Van Vigyan, *Journal of the Society of Indian Forester,* Vol. 20, 1982- Experimental sucker plantation of *Diospyros melanoxylon*.

Vaishampayan, S.M. and Bahadur, A (1986), Proceeding of Second Forestry conference, Vol.-II, 872-876 Government of India.

Vinod Kumar (1999), Nursery and Plantation Practices in Forestry– Scientific Publishers- 5-A. New Pali Road- Jodhpur.

Index